The Wingate Anaerobic Test

Omri Inbar, EdD
Wingate Institute

Oded Bar-Or, MD
McMaster University

James S. Skinner, PhD
Indiana University

Human Kinetics

aloging-in-Publication Data

The Wingate Anaerobic Test / Omri Inbar, Oded Bar-Or, James S. Skinner.
    p.   cm.
    Includes bibliographical references and index.
    ISBN 0-87322-946-0
    1. Exercise tests.   2. Anaerobiosis.   I. Bar-Or, Oded.
II. Skinner, James S., 1936-
QP303.I46   1996
613.7' 1'0287--dc20

96-12409
CIP

ISBN: 0-87322-946-0

Acquisitions Editor: Richard A. Washburn; **Developmental Editor:** Marni Basic; **Assistant Editors:** Susan Moore, Lynn M. Hooper, and Dawn Cassady; **Editorial Assistant:** Amy Carnes; **Copyeditor:** Elaine Otto; **Proofreader:** Pam Johnson; **Indexer:** Joan Griffitts; **Typesetter and Text Layout Artist:** Ruby Zimmerman; **Text Designer:** Judy Henderson; **Cover Designer:** Jack Davis; **Illustrator:** Studio 2-D; **Printer:** Versa Press

Printed in the United States of America   10 9 8 7 6 5 4 3 2 1

Human Kinetics
Web site: http://www.humankinetics.com/

United States: Human Kinetics
P.O. Box 5076, Champaign, IL 61825-5076
1-800-747-4457
e-mail: humank@hkusa.com

Canada: Human Kinetics, Box 24040, Windsor, ON N8Y 4Y9
1-800-465-7301 (in Canada only)
e-mail: humank@hkcanada.com

Europe: Human Kinetics, P.O. Box IW14, Leeds LS16 6TR, United Kingdom
(44) 1132 781708
e-mail: humank@hkeurope.com

Australia: Human Kinetics, 57A Price Avenue, Lower Mitcham, South Australia 5062
(08) 277 1555
e-mail: humank@hkaustralia.com

New Zealand: Human Kinetics, P.O. Box 105-231, Auckland 1
(09) 523 3462
e-mail: humank@hknewz.com

Contents

Preface

The Wingate Anaerobic Test (WAnT) is the most widely used anaerobic performance test, and a great deal of research has been done using it. Although many people use the WAnT, there has been no single book where they can find information about many aspects of the test. As a result, there is often confusion about the correct protocol and interpretation; this has led to variations in, for example, protocols, ergometers, and sophistication of recording devices, as well as in the comparability of results obtained. We noted a strong need for a book that discussed protocols, factors that might affect results, and the interpretation and use of data obtained from the WAnT.

This is a state-of-the-art review. We offer practical information about the WAnT and its administration, and we reexamine our previous research. Because we have been responsible for most of the research, we hope that this will be considered the authoritative work on the subject.

In this book we consider equipment and protocols and discuss feasibility, reliability, validity, and sensitivity. We then provide the reader with typical findings generated from the anaerobic exercise performance of children, adults, the elderly, athletes, and individuals with neuromuscular disabilities. Finally, we discuss how the test and its principles may be applied to other situations. We also offer suggestions for further research on testing muscle power, muscle endurance, and muscle fatigability.

This book should be useful for researchers and practitioners working in exercise physiology, physical education, physical therapy, and medicine (e.g., physical medicine and rehabilitation, pediatrics, pediatric cardiology), as well as for coaches and athletic trainers.

Acknowledgments

The Wingate Anaerobic Test was developed by members of the Department of Research and Sport Medicine at the Wingate Institute in Israel. The authors are particularly indebted to Raffy Dotan for his perseverance, creativity, and endless ideas. Alberto Ayalon conducted the first experiments intended to measure supramaximal muscle power on the cycle ergometer. Ira Jacobs and Jan Karlsson have performed validation studies in their laboratories. Information on the use of the test with the physically disabled, the elderly, and children with other diseases has been generated at the Children's Exercise and Nutrition Centre, McMaster University, Hamilton, Ontario, Canada. The authors are indebted to the Centre's staff and to Leslie McGillis in particular.

Credits

Figure 2.1
Reprinted, by permission, from H. Hebestreit, K. Mimura, and O. Bar-Or, 1993, "Recovery of anaerobic muscle power following 30-s supramaximal exercise: Comparing boys and men," *Journal of Applied Physiology* 74: 2875-2880.

Figures 2.3 and 2.4
Reprinted, by permission, from O. Inbar et al., 1983, "The effect of bicycle crank-length variation upon power performance," *Ergonomics* 26:1139-1146.

Figure 3.1
Reprinted, by permission, from E. Tirosh, P. Rosenbaum, and O. Bar-Or, 1990, "New muscle power test in neuromuscular disease. Feasibility and reliability," *American Journal of Diseases of Children* 144: 1083-1087. Copyright 1990, American Medical Association.

Figures 3.2 and 3.3
Reprinted, by permission, from O. Bar-Or et al., 1980, "Anaerobic capacity and muscle fiber type distribution in man," *International Journal of Sports Medicine* 1: 89-92.

Figures 3.4, 3.5, and 3.6
Reprinted, by permission, from C. Denis et al., 1992, "Power and metabolic responses during supramaximal exercise in 100-M and 800-M runners," *Scandinavian Journal of Medicine and Science in Sports* 2:62-69.

Figures 3.10 and 3.11
Reprinted, by permission, from O. Inbar and O. Bar-Or, 1980, Changes in arm and leg anaerobic performance in laboratory and field tests following vigorous physical training. In *Proceedings of the international seminar on the art and science of coaching* (Netanya, Israel: Wingate Institute), 38-48.

Figure 3.12
Reprinted, by permission, from C. Denis et al., 1990, Specific responses of the Wingate Anaerobic Test to sprint versus endurance training: Effects of the adjustment load. In *Proceedings of the Maccabiah—Wingate international congress, life sciences*, edited by G. Tenenbaum and D. Eiger (Netanya, Israel: Wingate Institute), 9-17.

Figure 4.1
Reprinted, by permission, from O. Inbar and O. Bar-Or, 1975, "The effects of intermittent warm-up on 7-9-year-old boys," *European Journal of Applied Physiology* 34: 81-89.

Figure 4.4
Reprinted, by permission, from O. Inbar et al., 1983, "The effects of alkaline treatments on short-term maximal exercise," *Journal of Sports Sciences* 1:95-104.

Figures 5.1 and 5.2
Reprinted, by permission, from O. Inbar and O. Bar-Or, 1986, "Anaerobic characteristics in male children and adolescents," *Medicine and Science in Sports and Exercise* 18: 264-269.

Figures 5.3 and 5.4
Reprinted, by permission, from C.J.R. Blimkie et al., 1988, "Anaerobic power of arms in teenage boys and girls: Relationship to lean tissue," *European Journal of Applied Physiology* 57: 677-683.

Figures 5.5, 5.6, 5.9, 5.10, 5.11, and 5.12
Adapted, by permission, from O. Inbar, 1985, *The Wingate Anaerobic Test. Its Performance, Characteristics, Application and Norms* (Hebrew) (Netanya, Israel: Wingate Institute).

Figure 5.7
Reprinted from E. Ben-Ari, O. Inbar, and O. Bar-Or, 1978, The anaerobic capacity and maximal anaerobic power of 30-40 year old men and women. In *Proceedings of the 5th international symposium of kinanthropometry and ergometry* (Quebec: Pelican), 427-433.

Figure 5.8, Table 2.1, Table 3.1, Table 3.2, and Table 3.4
Reprinted, by permission, from O. Bar-Or, 1987, "The Wingate Anaerobic Test—An update on methodology, reliability and validity," *Sports Medicine* 4: 381-394.

Figures 5.13 and 5.14
Reprinted, by permission, from O. Inbar, D.X. Alvarez, and M.A. Lyons, 1981, "Exercise-induced asthma—A comparison between two modes of exercise stress," *European Journal of Respiratory Disease* 62: 160-167.

Figure 5.15
Reprinted, by permission, from O. Bar-Or, 1986, "Pathophysiological factors which limit the exercise capacity of the sick child," *Medicine and Science in Sports and Exercise* 18: 276-282.

Table 4.1
Reprinted, by permission, from J.S. Skinner and D.W. Morgan, 1985, *Limits of human performance* (Champaign, IL: Human Kinetics), 40.

Table 5.1
Reprinted, by permission, from O. Bar-Or, 1983, *Pediatric sports medicine for the practitioner*, 13.

Chapter 1

Development of the Wingate Anaerobic Test

The Wingate Anaerobic Test (WAnT) was developed during the 1970s at the Department of Research and Sport Medicine of the Wingate Institute for Physical Education and Sport in Israel. Since the introduction of its prototype (Ayalon, Inbar, and Bar-Or 1974), the WAnT has been accepted in laboratories around the world to assess muscle power, muscle endurance, and fatigability (e.g., Bouchard et al. 1991). It has also been used as a reproducible, standardized task that can help analyze physiologic and cognitive responses to supramaximal exercise.

Historical Perspective on Anaerobic Test Development

The impetus for the development of the WAnT was the prevailing lack of interest in anaerobic performance as a component of fitness. Even today, many fitness appraisers, health professionals, and teachers consider fitness and physical working capacity to be synonymous with aerobic fitness. Numerous fitness surveys, for example, have focused on maximal aerobic power, completely ignoring such items as peak muscle power and local muscle endurance, even though these fitness components are important for various populations and situations. As an example, there are many daily and athletic events where it is essential to develop high-intensity power instantaneously or within a few seconds.

A major reason for this relative lack of interest in anaerobic exercise performance was the scarcity of appropriate, easily administered laboratory tests. No anaerobic test has reached the worldwide popularity of such aerobic protocols as the W170 or the Åstrand-Ryhming nomogram for predicting $\dot{V}O_2$max (maximal oxygen intake). By the beginning of the 1970s, few laboratory tests were available to assess muscle power, muscle endurance, and fatigability. The

1

oldest published test for short-term muscle thrust was the vertical jump (Sargent 1921). Although the variable measured in this test was distance, one could also calculate mechanical work. Power could not be calculated, however, because the time that the force acted on the body was not known. Calculation of power in the vertical jump became possible only by using a force platform (Davies and Rennie 1968). In 1971 Davies reported values of mechanical power during the vertical jump that were higher than ever described for any other task. Perhaps the best-known test of peak muscle power was the Margaria step-running test (Margaria, Aghemo, and Rovelli 1966), which allowed the calculation of peak mechanical power over less than 1 s. This test was used extensively by its developers and by others to study muscle energetics during supramaximal, short-term exercise.

Cycling protocols that were intended to measure or reflect "anaerobic capacity" or "anaerobic work" included the following: a 30- to 60-s test by De Bruyn-Prévost (1975); a similar protocol described by Chaloupecky (1972) in which subjects pedaled 85 revolutions at maximal speed against a constant (4 kp) resistance; and a 60-s test by Szögÿ and Cherébetiu (1974). Borg (1962) was interested in perceptual and motivational aspects of high-intensity exercise and developed a protocol of repeated 45-s tasks in which resistance was continuously increased at predetermined rates. Borg, Edstrom, and Marklund (1971) subsequently constructed a special ergometer for this "cycling strength and endurance test." At approximately the same time that the WAnT was developed, Katch (1973) used a 1-min supramaximal cycling task to analyze the kinetics of $\dot{V}O_2$max. Katch et al. (1977) later suggested a 40-s cycling test to analyze anaerobic power and anaerobic work.

Some authors have used a high-velocity treadmill run to exhaustion as an index of anaerobic performance (e.g., Cunningham and Faulkner 1969), but this approach did not yield a measurement of mechanical power. Furthermore, the endpoint was often determined by the subject's level of fear, rather than by physiologic causes; this subjectivity in response precluded the use of the treadmill test with disabled, elderly, and nonathletic populations.

Other laboratory measurements, such as oxygen deficit during intense exercise, oxygen debt following such exercise, and maximal blood lactate, have been used extensively to study anaerobic performance. While historically important for the study of muscle energy metabolism (e.g., Hermansen 1969; Hill, Long, and Lupton 1924; Margaria, Edwards, and Dill 1933), these variables were not specific enough to reflect performance during short-term, high-intensity exercise.

In the early 1970s, with the advent of the muscle biopsy, changes in such substrates as muscle ATP, CP, glycogen, and lactate were measured to analyze anaerobic metabolism (e.g., Bergström et al. 1971; Karlsson and Saltin 1970; Saltin and Karlsson 1971). Although these measurements were needed for our current understanding of muscle metabolism, they could not be adopted as routine fitness tests.

In 1972, Gordon Cumming of Canada presented a paper at the Wingate Institute, in which he described a 30-s all-out cycling test he had administered to children and adolescents within a battery of performance tests (Cumming 1973). That paper offered an important impetus for the Wingate team to launch its "anaerobic test project." The aims of this project were to develop and standardize a supramaximal test that would do the following:

- Measure muscle power rather than assess it indirectly through biochemical, physiological, or histochemical indices.
- Provide information on peak power, muscle endurance, and muscle fatigability.
- Be simple, i.e., it would require equipment that was commonly available and could be administered by personnel who did not require special skills and training.
- Be inexpensive, i.e., affordable to laboratories worldwide and to fitness appraisers who did not have the backing of a sophisticated laboratory.
- Be feasible, such that it could be performed by able-bodied and disabled people, by a wide spectrum of ages and fitness levels, and by either sex.
- Be a safe, noninvasive procedure that would be socially acceptable to various age and ethnic groups.
- Be easily performed by the upper or lower limbs, as anaerobic performance is a local, rather than a systemic, characteristic.
- Be objective, i.e., a subject's score would not depend on the observer's subjective interpretation.
- Be highly reliable and repeatable, i.e., the score would reflect the subject's actual performance, rather than a random occurrence that might change from one measurement to another.
- Be valid, i.e., the score would reflect the subject's supramaximal anaerobic performance capacity.
- Be sensitive, i.e., it would reflect changes over time (improvement and deterioration) in the subject's anaerobic performance.
- Be specific to anaerobic muscle performance rather than to fitness in general.

The new test was not intended to replace existing physiological, biochemical, or histochemical analyses of anaerobic muscle metabolism or to address basic issues of muscle contractility and fatigability. These would often require an invasive approach and equipment that is much more sophisticated and complex than that needed for the WAnT.

Characteristics of Anaerobic Performance Tests

Tests of anaerobic ability involve very high-intensity exercise lasting between a fraction of a second and several minutes (Skinner and Morgan 1985). It is generally agreed that most anaerobic tests are reliable in motivated subjects and that they correlate highly with each other, but there is less agreement about what they measure. There is also no "gold standard" with which to compare test results and, unlike the steady-state $\dot{V}O_2$ in submaximal tests or the plateau in $\dot{V}O_2$max seen in maximal aerobic tests, there are no accepted criteria for what should be measured or how.

This lack of agreement is partly caused by terminology. For example, the terms *anaerobic power* and *anaerobic capacity* have been used to describe the highest 3-s to 5-s and average 30-s power outputs, respectively, on the WAnT. Other anaerobic tests lasting 1-10 s (Komi et al. 1977; Margaria, Aghemo, and Rovelli 1966; Sargeant, Hoinville, and Young 1981) also use the term *anaerobic power*, while those lasting longer are often called tests of *anaerobic capacity* (Szögÿ and Cherébetiu 1974; Volkov, Shirkovets, and Borilkevich 1975). As a result, there are often discussions about the terminology itself and whether it adequately describes what is being measured. In this book, the highest 3-s to 5-s power output on the WAnT is operationally defined as peak power (PP) and the average 30-s power output is the mean power (MP). Reducing the use of labels should permit more discussion of the measures themselves and what they might mean.

It is difficult to determine the amount of aerobic and anaerobic involvement in tests lasting more than a few seconds. Estimates of relative contributions are questionable because of the difficulty in assessing mechanical efficiency of supramaximal tests (Vandewalle, Peres, and Monod 1987) and thus the total energy required. Given the large individual differences in (1) mechanical efficiency (Hermansen and Medbø 1984), (2) $\dot{V}O_2$max, and (3) the kinetics of the rise in $\dot{V}O_2$ at the onset of exercise, estimates of aerobic energy production and its relative contribution can vary greatly.

Along with the lack of accepted criteria of what constitutes the various types of anaerobic ability, there are big differences in the sophistication of measuring devices. As a result, differences in measurement accuracy and the frequency with which they are taken often make it difficult to compare results on the same anaerobic test obtained from different laboratories.

Types of Anaerobic Tests

There are many anaerobic tests. Based on their names and descriptions, however, there is not much agreement on what they measure. Nevertheless, because these anaerobic tests differ in intensity and duration, they can be classified as either very brief tests (lasting 1-10 s) or brief tests (lasting 20-60 s).

Very Brief Tests

Table 1.1 lists tests lasting 1-10 s. These involve running up stairs (Margaria, Aghemo, and Rovelli 1966), jumping (Davies 1971; Fox and Mathews 1974), contracting large muscle groups at maximal velocities (Komi et al. 1977; Thorstensson et al. 1976), short sprints (Fox and Mathews 1974), or pedaling (fastest 4-5 s in a test lasting 30-40 s) against a high resistance on a cycle ergometer (Bar-Or 1983; Crielaard and Pirnay 1981; Weltman, Moffatt, and Stamford 1978). The peak 3-s or 5-s power output of the WAnT is an example of the information available from this last type of test.

Table 1.1   Very Brief (1-10 s) Anaerobic Tests

Test	Approximate duration (s)
Margaria step test	2-4
Vertical jump	< 1
Leg extensor force	1-2
Short sprints	3-10
Cycle ergometer (max RPM)	
Resistance:[a]	
4-7 kg	2-5
4 kg	4-s max (40-s test)
75 gm · kg^{-1} body mass	5-s max (30-s test)

[a]These values are true for the Monark (or any other ergometer where the flywheel moves 6 m per pedal revolution) but not for the Fleisch ergometer.

Brief Tests

Examples of tests lasting 20-60 s can be seen in table 1.2. With the exception of long running sprints (Fox and Mathews 1974), which are field tests requiring little equipment, most tests use treadmills and cycle ergometers. Measurements in these tests include (a) total power output on the cycle ergometer within a specified time period (Szögÿ and Cherébetiu 1974) or (b) time to exhaustion against a given resistance on the cycle (De Bruyn-Prévost and Lefèbvre 1980) or at a given speed and grade on the treadmill (Cunningham and Faulkner 1969; Houston and Thomson 1977; Roberts, Billeter, and Howald 1982; Sjödin et al. 1976). The WAnT is a good example of the former type, as it measures the mean power output (MP) of the arms or legs during 30 s. Campbell et al. (1979) determined the total pedal revolutions during two 20-s tests at power outputs requiring 75% and 150% $\dot{V}O_2$max, while Katch et al. (1977) and Weltman, Moffatt, and Stamford (1978) did the same during 30-40 s against a resistance of 4-6 kilograms.

Table 1.2   Brief (20-60 s) Anaerobic Tests

Test	Duration (s)
Cycle ergometer (arms)	
50 gm · kg^{-1} body mass at max RPM[a]	30
Cycle ergometer (legs)	
To exhaustion	
350-400 W at 104-128 RPM	40-45
Fixed duration	
1.5 · $\dot{V}O_2$max at max RPM	20
0.75 · $\dot{V}O_2$max at max RPM	20
75 gm · kg^{-1} body mass at max RPM[a]	30
4-6 kg at max RPM[a]	30-40
Treadmill run to exhaustion	
7-8 MPH at 20% grade	30-60
10 MPH at 15% grade	35-45
Individually designed to elicit exhaustion in 45 s	35-60

[a]Resistance values are true for the Monark (or any other ergometer where the flywheel moves 6 m per pedal revolution) but not for the Fleisch ergometer.

The Wingate Anaerobic Test in Perspective

We are often asked whether the WAnT is the best anaerobic power test. Our answer is that there is insufficient information to determine whether any given test is superior to others, but the WAnT is "the most-tested test." We are not aware of other tests for peak muscle power or for muscle endurance that have been evaluated in the literature as thoroughly as the WAnT. Numerous laboratories have confirmed its very high reliability as well as its validity as a test that can yield information on peak mechanical power and on local muscle endurance.

We cannot tell definitively whether a 30-s protocol (as used in the WAnT) is superior to a 20-s or 45-s protocol. Observations made by the Wingate group in the mid-1970s suggested that scores in all-out sprint cycling tests lasting 15, 30, or 45 s were very highly correlated. Indeed, a 45-s protocol can elicit more mechanical work than does the WAnT 30-s protocol and, as such, it can bring subjects closer to their capacity. However, subjects who perform a 45-s protocol are often reluctant to ever perform it again. This makes a 45-s protocol less suitable for studies of athlete/patient evaluations that require repeated testing.

One disadvantage of the WAnT is that it measures performance of several muscle groups combined and therefore cannot yield information about any specific muscle or muscle group. If such information is sought (e.g., in studies of muscle contractility or for the generation of force-velocity curves), then one should use such dynamometer-based monarticular tests as knee extension or knee flexion.

Chapter 2

Description of the Wingate Anaerobic Test

This chapter describes the WAnT protocol, equipment needed, measures obtained, and ways to standardize the test. As in many new tests that have evolved gradually, the WAnT of today is slightly different from that presented in the initial publications. Most of the research and findings presented here are based on the final standardized version.

Protocol

The WAnT requires pedaling or arm cranking for 30 s at maximal speed against a constant force. This force is predetermined to yield a supramaximal mechanical power (usually equivalent to about two to four times that associated with the maximal aerobic power, or $\dot{V}O_2$max) and to induce a noticeable development of fatigue (i.e., a drop in mechanical power) within the first few seconds.

Equipment Needed

In its simplest form, the WAnT can be administered using only a mechanical ergometer (e.g., Fleisch, Monark, or Bodyguard) and a stopwatch. Pedal revolutions can be counted visually. More sophisticated methods also have been used and are discussed later in this chapter.

Warm-Up

A 5-min to 10-min intermittent warm-up (alternating 30-s exercise with 30-s rest) was found best suited for optimal performance on the WAnT (Inbar and Bar-Or 1975). The warm-up should be done on a cycle ergometer to promote more specific physiological and motor adaptations. Mean heart rate should be about 160 beats · min^{-1} in children and 150 beats · min^{-1} in young adults. A shorter version has also been used when needed and consists of 2-4 min of pedaling interspersed with two or three "all-out sprints" each lasting 4-8 s so that subjects can get a feel for the actual test. Subjects should then rest 3-5 min to eliminate any fatigue associated with the warm-up. Even though various versions for warming up have not been compared systematically, the procedures mentioned above have been found useful for various groups, irrespective of age, gender, or health status.

Testing

With the command "start," subjects pedal as fast as possible against a low resistance to overcome the inertial and frictional resistance of the flywheel and to shorten the acceleration phase. If possible, the investigator should make certain that the maximal speed has been reached because subjects cannot speed up once the full predetermined load has been applied; this phase normally takes 3-4 s. The full load is then applied to start the 30-s test. If it is not possible to have a direct measure of the revolutions before the test begins, subjects can say "go" when they feel that they are at their maximal speed. Some laboratories have developed customized software that puts the revolution counter on-line so that the program can determine when the full resistance can be applied. As soon as the resistance is applied, the counting of revolutions begins and lasts exactly 30 s. Verbal encouragement should be given throughout the test, especially during the last 10-15 s when discomfort is greatest and more willpower is needed.

When explaining the WAnT to subjects, the tester must emphasize the need to pedal as fast as possible from the beginning and to maintain maximal speed throughout the 30-s period. Any attempt to "conserve energy" for the last few seconds will give inadequate results and should be strongly discouraged. The test is a maximal 30-s effort, and nothing less should be accepted.

Cooldown

All subjects, especially adults and tall adolescents, should be regarded as being potentially susceptible to dizziness and syncope following the exertion of the WAnT. To prevent these potential problems, 2-3 min of pedaling against a light resistance immediately following the test is strongly recommended.

Repeated Testing

As Hebestreit, Mimura, and Bar-Or (1993) have shown, the performance of most young adult men recovers fully within 10 min after they have completed the WAnT. Prepubescent boys, on the other hand, fully recover within 2 min (fig. 2.1). Therefore, to guarantee complete recovery before repeating a test, a rest interval of at least 20 min is recommended when retesting the same muscle group. When testing the arms after the legs, a 15-min interval is possible. When testing the legs after the arms, a shorter interval is usually adequate.

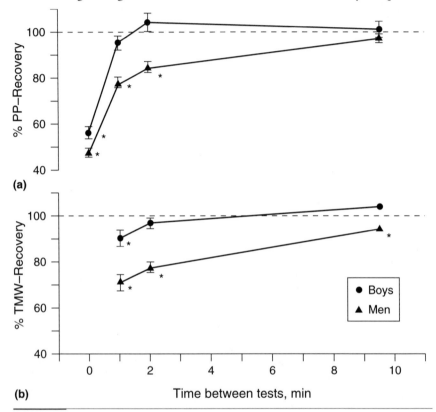

Figure 2.1   Relationship between % recovery and recovery time in eight boys and eight men. (a) Recovery of peak power (PP). $F_{boys/men}$ = 23.2 ($p < 0.001$); $F_{recovery\ time}$ = 9.6 ($p<0.001$). (b) Recovery of total mechanical work (TMW). $F_{boys/men}$ = 35.3 ($p < 0.001$); $F_{recovery\ time}$ = 52.5 ($p < 0.001$). *WAnT2 significantly different from WAnT1 ($p < 0.05$). Note: % recovery of PP at recovery *time 0* was estimated from power output during last 6 s of WAnT1. This hypothetical value will be slightly higher than the real value and has not been used for statistics.
Reprinted from Hebestreit, Mimura, and Bar-Or 1993.

Indices of Performance

Figure 2.2 shows a typical curve using 3-sec segments and plotting the average value at the midpoint of each segment. While there might be theoretical and practical reasons for measuring instantaneous mechanical power throughout each pedal revolution, the WAnT looks at pedal revolutions against a constant resistance as its measure of mechanical power. Because the amount of time required to complete each revolution varies, the WAnT uses the average of 3-s to 5-s segments. The choice of segment duration has little or no effect on the results.

As shown in figure 2.2, three indices are measured:

- Peak power (PP) is the highest mechanical power elicited during the test, as shown by point A, which typically occurs in the first few seconds; this index is usually taken as the average power over any 3-s or 5-s period.

- Mean power (MP) is the average power sustained throughout the 30-s period. It is obtained by averaging the values obtained during the ten 3-s or six 5-s segments.

- Rate of fatigue (Fatigue index or FI) is the degree of power drop-off during the test. As shown in figure 2.2, this is calculated as a percentage of PP. It can also be taken as the slope of the straight line connecting points A and B (the lowest power), i.e., A – B, divided by the time from the start of the test until point B is reached.

NOTE: Some laboratories calculate "total work" instead of MP. Total work is the product of MP and time. As is the case with maximal aerobic power ($\dot{V}O_2$max) and for similar reasons, PP and MP can be expressed in absolute values (e.g., Watt), relative to body mass (e.g., $W \cdot kg^{-1}$) or relative to lean body mass (e.g., $W \cdot kgLBM^{-1}$).

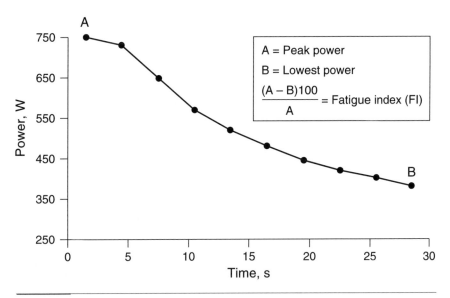

Figure 2.2   The Wingate Anaerobic Test. A typical curve.

Most research so far has focused on the PP and MP as fitness indices. Much less is known about the relevance of the fatigue indices to anaerobic fitness. The fatigue indices are used less often because of their greater variability among subjects, as well as their close association with PP. Even though relatively strong associations have been reported between the FI and various histological measures of muscle fiber composition, testers should use caution when using FI to interpret specific performance and/or physiological characteristics of a given individual.

Originally, PP was assumed to reflect the alactic (phosphagen) anaerobic processes and MP was assumed to reflect the rate of anaerobic glycolysis in the muscle. A subsequent study (Jacobs et al. 1983) showed that muscle lactate rises to extremely high levels as early as 10 s into the test. Therefore, PP is unlikely to reflect alactic processes alone. In many publications, MP also has been called *anaerobic capacity*. This is based on an unproven assumption, and we prefer not to use this term. It is safe to assume, however, that PP is a reflection (although not a direct measurement) of the ability of the limb muscles to produce high mechanical power within a short time. On the other hand, MP reflects the local endurance of these muscles (their ability to sustain extremely high power).

Standardization

In order to have comparable results from one test to another and from one laboratory to another, the administration of the WAnT must be standardized. The remaining parts of this chapter outline some elements of the test that should be standardized.

Equipment

In the last decade, several laboratories have raised the sophistication of their ergometers and recording techniques. Mechanical ergometers based on hanging weights rather than a pendulum afford a more precise administration of the force. The Fleisch-Metabo ergometer (Basel, Switzerland), in which weight can be incremented by as little as 60 g, has been used in the laboratories at the Wingate Institute and McMaster University. Electrically braked ergometers are suitable only if they have a constant-force mode, e.g., Siemens.

Automatization of the counting has also increased precision. Some laboratories have been using a custom-made set of counters. Others have connected the "standard" channel of a strip-chart recorder (e.g., an electrocardiograph) to a sensor that is tripped by the pedal crank. Pedal counts sensed by electromagnets or photoelectric cells also can be fed into a microcomputer, which generates an on-line analysis of power and can calculate all indices.

Typical cycle ergometers used for the Wingate Test. The ergometer on the right (Fleisch-Metabo) is identical to the one used at the Wingate Institute for developing the test. It is mechanically braked, with small weights added to increase the friction between the belt and flywheel. The ergometer on the left (WatSystem) is electronically controlled, and it represents a line of ergometers developed specifically for the Wingate Test.

Pedal Crank Length. The conventional length of the pedal crank in ergometers is 17.5 cm. In most laboratories, this length is used for all subjects, irrespective of height or leg length. Theoretically, however, optimal crank length should vary with the leg (or arm) length of the individual, whether the task is aerobic or anaerobic. Although lack of such optimization may affect several variables, e.g., angle of application, torque, kinetic energy (and energy loss) of moving the leg mass, and muscle tension and force-velocity relationships, changing crank length appears to have little effect on the performance of the WAnT. Inbar et al. (1983a) had thirteen male nonathletes, ages 22 to 27, perform the WAnT at crank lengths of 12.5, 15.0, 17.5, 20.0, and 22.5 cm. As there was a great deal of overlap in the curves, only the results for crank lengths of 15 and 22.5 cm are shown in figure 2.3. Using best-fit parabolic curves, the optimal calculated crank lengths were 16.4 cm for MP and 16.6 cm for PP, respectively (fig. 2.4). It was further found that the optimal crank length depended on the subject's lower limb length. A limb-to-crank ratio of 6.3 seemed optimal in these young adults whose limb lengths ranged from 92 to 107.2 cm. No data are available for people who are taller or shorter. It should be emphasized, however, that a deviation from the optimal length by as much as 5 cm affected MP by only 0.77% and PP by only 1.24%. The practical importance of using crank lengths other than 17.5 cm is therefore marginal, unless one studies children or extremely tall adults.

A pedal whose crank-arm length is adjustable to one of three positions (it can be shortened for testing children and for upper-limb ergometry). The extra holes, as shown in the picture, can be drilled and threaded by a local machinist.

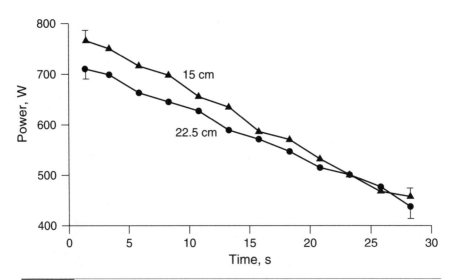

Figure 2.3   Changes in power output during a 30-s supramaximal effort at 15-cm and 22.5-cm crank lengths ($\bar{X} \pm SEM$).
Reprinted from Inbar et al. 1983.

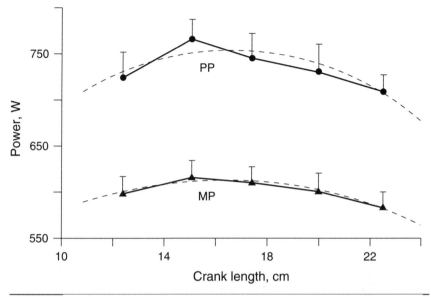

Figure 2.4   Mean (MP) and peak (PP) power output as a function of crank length ($X \pm SEM$). Dashed lines represent best-fit parabolic curves. The r_{parab} for MP and PP was 0.98 and 0.87, respectively.
Reprinted from Inbar et al. 1983.

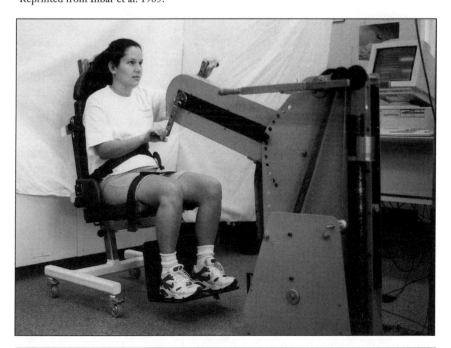

A subject using the Fleisch-Metabo ergometer to perform the upper-limb version of the Wingate Test. The height of the crank arm axis is positioned at shoulder level. In this example the ergometer is connected on-line to a personal computer.

Use of Toe Stirrups.   Mounting toe stirrups on the pedals represents an important methodological refinement. Lavoie et al. (1984) found that stirrups increased PP and MP of young adult athletes and physical education students by 5% to 12%. The apparent reason is that a pushing or pulling force can be exerted on the pedal throughout the full cycle. Therefore, using any system that allows the application throughout the cycle is recommended with all subjects, including nonathletic children and adolescents. For arm tests, a cylindrical pedal with no additional attachments is adequate because the grasp enables one to exert force throughout the cycle.

The use of toe clips enables the user to exert force on the pedal during both knee flexion and knee extension.

Optimization of Resistance

Choosing a force setting that would elicit the highest possible PP and MP for each subject is important and, as yet, is only a partially resolved challenge. The force originally suggested by the Wingate group was 0.075 kp per kg body mass (assuming the use of a Monark ergometer). This force is equivalent to a mechanical work of 4.41 Joule per pedal revolution per kg body mass. The choice of this force was based on a study of a small group of young untrained individuals and, in retrospect, appears to be too low for most adults.

Indeed, as summarized in table 2.1, subsequent reports have shown that the optimal force is higher than originally suggested. Evans and Quinney (1981) studied male physical education students and varsity athletes using a modified

Monark ergometer. The resistance (group average) that yielded the highest MP was 0.098 kp · kg⁻¹, which is equivalent to 5.76 Joule per pedal revolution per kg body mass. The authors suggested the following equation for calculating the optimal force for an individual based on body mass and leg volume (as determined by a water displacement technique):

$$\text{Force (kp)} = -0.4914 - 0.2151 \text{ (Body Mass in kg)} + 2.1124 \text{ (Leg Volume in L)}$$

This equation was tried with a group of nonathletic military personnel and found to have a low validity (Patton, Murphy, and Frederick 1985). In another study by Lavoie et al. (1984), force was applied according to this equation. They found a higher PP but a similar MP compared with values obtained using a force of 0.075 kp · kg⁻¹. The need to measure leg volume introduces an added requirement that might interfere with the test's feasibility in some laboratories.

Using a Fleisch ergometer, Dotan and Bar-Or (1983) tested 18 female and 17 male physical education students and found an inverted-U relationship between force and MP. As seen in figure 2.5, the force that yielded the highest MP for the males was 52 g · kg⁻¹ (equivalent to 0.087 kp · kg⁻¹ on the Monark

Table 2.1   The Optimal Load for Yielding the Highest Mean Power, Based on the Prototype Wingate Anaerobic Test and on Subsequent Studies

Subjects	Limb	Force (kp · kg⁻¹) Monark	Fleisch	Work (J · rev⁻¹ · kg⁻¹)	References
Adult males					
Sedentary	Legs	0.075	0.045	4.41	Ayalon et al. (1974)
Active and athletes	Legs	0.098	0.059	5.76	Evans and Quinney (1981)
Physical education students	Legs	0.087	0.052	5.13	Dotan and Bar-Or (1983)
Soldiers	Legs	0.094	0.056	5.53	Patton et al. (1985)
Physical education students	Arms	0.062	0.037	3.62	Dotan and Bar-Or (1983)
Adult females					
Physical education students	Legs	0.085	0.051	5.04	Dotan and Bar-Or (1983)
Physical education students	Arms	0.048	0.029	2.82	Dotan and Bar-Or (1983)
Boys (13-14 years)					
Active, nonathlete	Legs	0.070	0.042	4.13	Dotan and Bar-Or (1983)
Girls (13-14 years)					
Active, nonathlete	Legs	0.067	0.040	3.92	Dotan and Bar-Or (1983)

Reprinted from Bar-Or 1987.

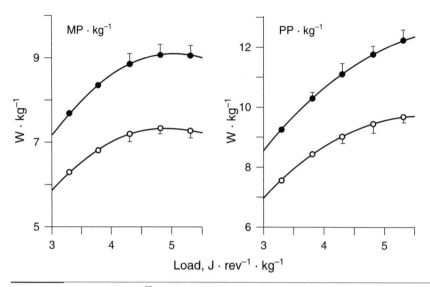

Figure 2.5   Mean values ($\overline{X} \pm SEM$) of mean and peak power outputs normalized for body mass at increasing ergometer leg-test loads in men (●) and women (○). Best-fit parabolic curves have been applied to the MP · kg⁻¹ data. The curves' maxima define the respective load optima.

ergometer). The respective force for the females was just a little lower. The intergender difference in optimal force was larger for arm exercise (fig. 2.6). As can be seen in figure 2.5, however, the PP did not appear to plateau and probably occurred at a higher force. Thus, it is unlikely that a single test can be optimized for both MP and PP.

In research done at Arizona State University, Heiser (1989) studied 8 endurance athletes, 4 power athletes and 10 untrained men on a Monark ergometer using force settings ranging from 0.075 to 0.105 kp · kg⁻¹. All three groups had a rise in PP with increasing force, suggesting that the optimal force was greater than 0.105 kp · kg⁻¹ (fig. 2.7). Although the lowest MP coincided with 0.075 kp · kg⁻¹ and the highest MP occurred at 0.105 kp · kg⁻¹ for all three groups, the rise in MP of about 1.1 to 1.3 W · kg⁻¹ was not significant due to the large interindividual differences (fig. 2.8).

In a study by Patton, Murphy, and Frederick (1985) of young military personnel who were tested with a Monark ergometer, the optimal force for MP was 0.094 kp · kg⁻¹, which is equivalent to 5.53 Joule per pedal revolution per kg. In this study and in the one by Dotan and Bar-Or (1983), optimal force for PP was higher than that needed to maximize MP.

To define the optimal force in their 40-s all-out cycling test, Katch et al. (1977) compared the power output of 30 young males at three force settings on the Monark ergometer: 4.0, 5.0, and 6.0 kp (corresponding to average relative forces of 0.053, 0.067, and 0.080 kp · kg⁻¹, respectively). The latter two loads elicited a higher total work output than 4.0 kp, but they did not differ

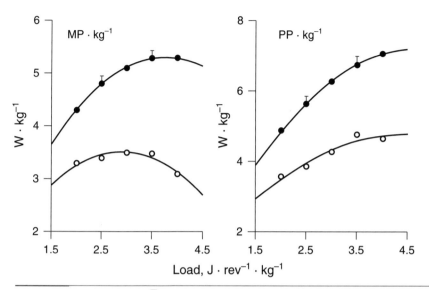

Figure 2.6   Mean values ($\overline{X} \pm SEM$) of mean and peak power outputs normalized for body mass at increasing ergometer arm-test loads in men (●) and women (○). Best-fit parabolic curves have been applied to the MP · kg^{-1} data. The curves' maxima define the respective load optima.

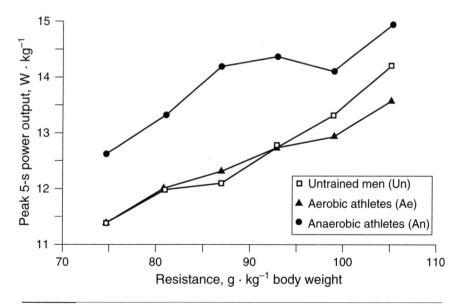

Figure 2.7   Peak 5-s power output (W · kg^{-1}) on the Monark ergometer for 10 untrained men (Un), 8 aerobic athletes (Ae), and 4 anaerobic athletes (An) on the Wingate Anaerobic Test.

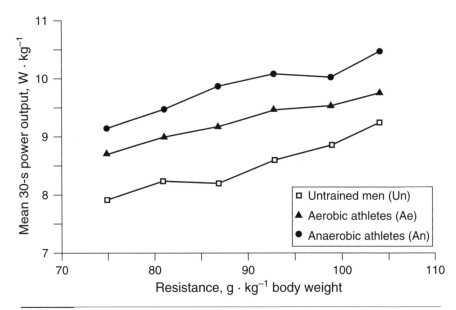

Figure 2.8   Mean 30-s power output (W · kg⁻¹) on the Monark ergometer for 10 untrained men (Un), 8 aerobic athletes (Ae), and 4 anaerobic athletes (An) on the Wingate Anaerobic Test.

from each other. Thus, the authors recommended a force of 5.0 to 6.0 kp for that test, which is lower than that recommended for the WAnT; this difference probably reflects the longer duration of the Katch test.

Based on force-velocity and power-velocity relationships, as developed from isokinetic cycling, the cranking velocity yielding the highest average power during 30 s is approximately 100 to 110 rpm (McCartney, Heigenhauser, and Jones 1983; Sargeant 1989). This is in agreement with a pedaling rate of 108 rpm (100 rpm in children) found when the subjects of Dotan and Bar-Or pedaled against optimal force. For unknown reasons, pedaling velocity at optimal force was some 20% lower in females than in males and some 12% to 15% lower in arm than in leg exercise (Dotan and Bar-Or 1983).

An alternative approach for optimizing the braking force is based on the Force-Velocity Test (Pirnay and Crielaard 1979; Sargeant, Hoinville, and Young 1981). In this test, which is intended to find a person's peak mechanical power, the subject usually performs five to eight all-out pedaling or arm-cranking sprints of 5-7 s. Each sprint is performed against a different braking force and yields a different power. The highest power is then considered the person's peak power. It has been suggested (Van Praagh et al. 1990; Vandewalle et al. 1985) that the braking force that yields the peak power in the Force-Velocity Test can be used as the optimal force for the WAnT. This approach has been validated for children with neuromuscular disabilities (Van Mil et al. 1993), for whom the optimal braking force for the WAnT was 65% of that found in the Force-Velocity Test. For details, see chapter 5.

In conclusion, the force needed to yield the highest MP is some 20% to 30% higher than originally suggested and seems to depend on the training level of the subject, being highest among athletes, particularly those engaged in events that require high power. It is also higher for adults than for children and marginally higher for males than for females. The force needed to elicit the highest PP is higher than that needed for the highest MP. As a general guideline with the Monark ergometer, it is recommended that a force of 0.090 kp · kg⁻¹ be used with adult nonathletes and 0.100 kp · kg⁻¹ with adult athletes.

A word of caution is warranted, however. Heiser (1989) used a strain gauge to measure strain applied to the ergometer chain to test whether the force applied to the front-loading basket of the Monark ergometer corresponded to the force applied at the flywheel. Five endurance athletes and seven untrained subjects pedaled for 5 s against 5.5 to 10.5 kg in 0.5-kg increments at 120 and 90 rpm; these speeds were considered representative of those achieved during the middle and later stages of the WAnT, respectively. For both speeds and for both groups, there was a linear increase in force applied to the pedals up to a weight of about 9.5 kg, after which the values tended to level off (fig. 2.9).

The implication of these findings is that there are upper limits to the accuracy of the Monark ergometer. For example, using the recommended force of 0.090 kp · kg⁻¹, the heaviest adult nonathlete who can be accurately tested would be 105 kg. Similarly, using the recommended force of 0.100 kp · kg⁻¹, the heaviest adult athlete would be 95 kg. Given that many power athletes (e.g., football players, weight lifters) weigh more than 95 kg, using the Monark

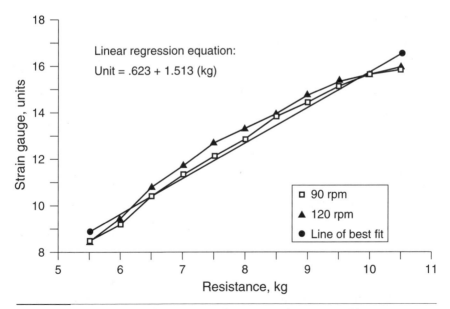

Figure 2.9 Mean values of strain gauge data at 90 and 120 rpm and linear regression line. Although the units are arbitrary and do not correspond to actual forces, the strain gauge readings were calibrated with several forces before each test.

ergometer for the WAnT is definitely limited. It is not known whether the accuracy of other ergometers is also reduced at higher forces. It should be emphasized that the Monark and other mechanical ergometers were designed only for aerobic testing but have been used at higher forces.

Conceptually, selecting the optimal force according to total body mass may not be the best approach (fat-free mass or muscle mass may be better alternatives). For practical purposes, however, the use of body mass as a criterion seems reasonable. One should also realize that selecting a force that is somewhat off the actual optimal force introduces only a small underestimate of the true MP (Dotan and Bar-Or 1983; Patton, Murphy, and Frederick 1985). For example, administering a force equivalent to 0.5 Joule per revolution per kg higher or lower than the real optimum underestimated MP by only 3% to 5% in males and females who performed either arm or leg tests (Dotan and Bar-Or 1983).

Data collected at the Children's Exercise and Nutrition Centre at McMaster University clearly show that, among individuals with disabilities, a choice of optimal force based on body size is meaningless. This is particularly true for children with severe muscular or nutritional disease, for whom the optimal load may be as low as 30% to 50% of that assumed from body mass.

Despite obvious logistical drawbacks, it seems that optimal load settings should eventually be done on an individual basis. Such a procedure could be based on relationships between velocity and force and/or power, as has been done for some time in several laboratories in France (Vandewalle et al. 1985).

Test Duration

The WAnT was originally structured using the 30-s cycling test described by Cumming (1973). A 30-s duration was also considered suitable for taxing anaerobic glycogenolysis, based on Margaria's findings with supramaximal treadmill runs to exhaustion (Margaria et al. 1969). A main consideration for choosing 30 s was based on pilot observations in which 30-, 45-, and 60-s protocols were compared. While all subjects managed to elicit an all-out effort throughout the 30-s test, some repeatedly tried to start at less than all-out speed in the longer tests because they were not certain they could complete the task. Such tactics would have elicited an underestimation of PP and might possibly detract from the reproducibility of the test. There is also an ever-increasing contribution of aerobic processes with the longer durations.

Katch et al. (1977) administered a 2-min all-out test to young adults. Based on statistical considerations, they concluded that a 40-s task would be more representative and they therefore recommended using a 40-s anaerobic test. Data comparing the PP and MP in the WAnT with those obtained in the 40-s test showed significantly lower values in the longer test (Maud and Schultz 1989). Furthermore, children, the elderly, or the disabled may not be able to complete a 40-s all-out task.

Because longer tests become more aerobic and less strenuous, Vandewalle, Peres, and Monod (1987) suggest that tests longer than 60 s are not needed, especially because Katch et al. (1977) found a correlation coefficient of 0.95 between total work output in a 40-s test and a 120-s test. Regardless of the resistance applied, Raveneau (unpublished thesis cited by Vandewalle, Peres, and Monod 1987) found that the correlation coefficient between total work done at 20 s and that done at 30 s of an all-out test was 0.99. Thus, Vandewalle, Peres, and Monod (1987) suggest that anaerobic tests need be only 15-20 s because they are easier to perform than tests lasting 30-40 s. For this reason, further studies are indicated to ascertain the optimal duration for an anaerobic-cycling and arm-cranking test that will yield meaningful data on PP, MP, total work, and fatigue with a minimal aerobic involvement. Nevertheless, because a vast amount of data have been generated using the 30-s test, we recommend that this duration be used for comparative purposes.

Safety Considerations

To examine possible health hazards associated with the WAnT performance, some subjective and physiological responses associated with well-being were tested before, during, and after performance of the WAnT with both the legs and arms (Inbar, Rotstein, and Dlin, unpublished data). This project was undertaken following the empirical observation that nausea and vomiting occurred in 5% of adults tested. In an attempt to determine the factors responsible for such feelings, 15 adults performed the test under the following four conditions with 2 days of rest between each treatment: (1) standard active warm-up and cooldown; (2) no warm-up but a cooldown; (3) no warm-up and no cooldown; and (4) warm-up but no cooldown. The cooldown consisted of continuous light pedaling against minimal resistance at a comfortable rate for 3-4 min.

Physiological indices studied before and after the test were blood pressure, ECG, pH, blood lactate, lactate dehydrogenase (LDH), and creatine phosphokinase (CPK). When the test was performed by either the legs or arms without an active cooldown, several physiological responses were observed that might explain, even among healthy people, the unpleasant side effects previously described. These include a sharp decrease in blood pH with a concomitant rise in lactic acid, a steep drop in blood pressure, and an increase in concentrations of LDH and CPK. A positive and significant correlation was found between those changes and the subjective feelings of the subjects at the end of the WAnT.

On the other hand, doing an active cooldown prevented those physiological responses and markedly improved the subjective feeling of the persons tested.

Thus, it is reasonable to assume that performing the test with a warm-up and/ or cooldown does not constitute a health hazard and can be safely used by healthy persons.

Other researchers at McMaster University (Jones and McCartney 1986; Markides et al. 1985) have used a 30-s isokinetic cycling test with healthy subjects up to 70 years of age and report no untoward effects. The same test was performed successfully by 33 middle-aged people with coronary artery disease (Jones and McCartney 1986), most of whom were tested within 3 months of sustaining a myocardial infarction and some following a coronary bypass operation.

These observations suggest that such brief, high-intensity tasks are safe with elderly people, including those with respiratory or coronary disease. Nevertheless, special caution must be taken with such subjects, including a thorough warm-up and continuous monitoring of the blood pressure before and after the test and the electrocardiogram before, during, and after the test.

Chapter 3

Characteristics of the Wingate Anaerobic Test

When deciding whether to use a particular written or physical test, one would like to know how reliable, valid, and sensitive it is. This chapter reviews the research on the WAnT for these critical characteristics.

Reliability

Table 3.1 summarizes studies of the WAnT's test-retest reliability. Correlation coefficients for tests performed under standardized environmental conditions have ranged between 0.89 and 0.99, but they are usually higher than 0.94. They tend to be somewhat higher for MP than for PP, which may reflect the higher error in the measurement of PP. Such high reliability coefficients are not found only with able-bodied individuals, however. In a study completed at McMaster University (Tirosh, Rosenbaum, and Bar-Or 1990), 58 children and adolescents with spastic cerebral palsy, athetotic cerebral palsy, muscular dystrophy, and muscular atrophy performed the arm-cranking WAnT in duplicate, 7-14 days apart. Test-retest reliability coefficients were 0.94 for PP and 0.98 for MP. Thirty-eight of these subjects also performed the leg WAnT test in duplicate, and in this case, reliability coefficients were 0.96 for both PP and MP. Figure 3.1 contains the individual performance data from both trials. Such high correlation coefficients are particularly impressive when one realizes that some children suffered from involuntary limb movement and their hands and feet had to be tied to the pedals. Another experiment was performed with 19 patients ages 54 to 84 years (mean age 67.6 years) with chronic obstructive lung disease. These men and women performed an abbreviated form of the leg WAnT (15 s) in duplicate, with 60 min separating the two trials. The test-retest coefficients were 0.89 for both PP and MP (Bar-Or, Berman, and Salsberg 1992). Whether done on the same day or several weeks apart, this test is highly reliable. In fact, Hebestreit, Mimura, and Bar-Or (1993) found

Table 3.1 Test-Retest Reliability (*r*) of the Wingate Anaerobic Test

Subjects	*r*	Comments	Reference
Children and young adults	0.95–0.97	Several experiments	Bar-Or et al. (1977)
Elderly, COPD patients (18)	0.89	Abbreviated WAnT	Berman and Bar-Or (unpublished)
Active or athletic young adults (12)	0.96		Evans and Quinney (1981)
Physical education students and athletes (9)	0.95–0.97		Kaczkowski et al. (1982)
Girls and boys, 10- to 12-year-olds (28)	0.89–0.93	3 climates, 2-wk intervals	Dotan and Bar-Or (1983)
Military personnel (19)	0.91–0.93		Patton et al. (1985)
N-M disease, 6- to 20-year-olds (58)	0.94–0.98	Arm test	Tirosh et al. (1990)
N-M disease, 6- to 20-year-olds (38)	0.96	Leg test	Tirosh et al. (1990)
Boys, 8- to 12-year-olds (8)	0.95–0.98	Leg test	Hebestreit et al. (1993)
Men, 18- to 23-year-olds (8)	0.93–0.99	Leg test	Hebestreit et al. (1993)

Note: COPD = chronic obstructive pulmonary disease; N-M = neuromuscular.
Reprinted from Bar-Or 1987.

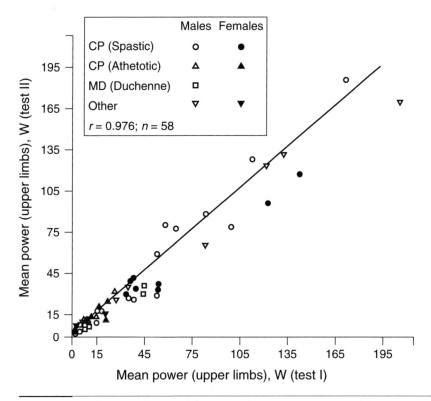

Figure 3.1   Reliability of the Wingate Anaerobic Test (mean power of the upper limbs on 2 d) with 58 disabled children and adolescents with spastic and athetotic cerebral palsy (CP), muscular dystrophy (MD, Duchenne), and other forms of neuromuscular disease (other).

Reprinted from Tirosh, Rosenbaum, and Bar-Or 1990.

that 20 min of rest between tests is more than adequate for reliable results; this is a major strength of the WAnT. Thus, recent data confirm initial observations that the WAnT is a highly reliable task.

Validity

To determine the validity of any test (i.e., how accurately it measures what it is meant to measure), one should compare it to an established "gold standard." The validity of the WAnT should therefore be determined against another test that has been chosen as the best test of anaerobic performance. When planning validation studies for the WAnT, the Wingate group was confronted with a problem. While several histologic and physiologic variables and existing tests reflected anaerobic performance, none could be considered a "gold standard" that measured both peak mechanical power and local endurance of the upper and lower limbs. A conceptually similar problem was expressed by Katch et al. (1977) when they developed their 40-s anaerobic cycling test. It was therefore decided to compare performance on the WAnT with several indices of anaerobic performance. The following sections describe such studies, as well as data generated in other laboratories.

Comparison With Anaerobic Performance in the Field

Table 3.2 summarizes studies in which WAnT indices were correlated with performance in sprinting, short-distance swimming, a short-term ice-skating task, and the vertical jump. All these tasks require very high-intensity exertion and last several seconds (the longest is the 300-m run, which lasts 50-70 s in children) and therefore can be considered predominantly "anaerobic" tasks. However, success in each is also skill dependent, and none can be taken as a "gold standard." Furthermore, standardization of these tasks is not usually as stringent as is possible in a laboratory setting.

Most of these observations were generated in seven laboratories and yielded correlation coefficients of 0.75 or more (equivalent to a common variance of 56% or more). The strongest association was found with the short sprint and 25-m swimming. The weakest association ($r = 0.32$) was with a shuttle ice-skating task (SAS40 = Sargeant Anaerobic Skating test), which apparently required a high skill level. Thus, it appears that the relation between WAnT power indices and "anaerobic" performance tasks is quite high, but not high enough to use the WAnT as a predictor of success in these specific tasks. It is probable, however, that MP and/or PP would be major predictors within a multiple-regression equation.

Skinner and O'Connor (1987) did a cross-sectional study on five groups of athletes who trained differently. There were two anaerobic groups (power lifters and gymnasts), two aerobic groups (10-km runners and ultramarathoners), and one mixed group (wrestlers). Power lifters had a significantly higher PP than ultramarathoners. The other athletes were distributed in a logical order, based on their type of training (table 3.3).

Interestingly, there were no differences in MP among the athletes. This was explained by the fact that the anaerobic athletes began with higher values but had a faster rate of fatigue, whereas the aerobic athletes had lower initial values and less of a decrement over 30 s. This is probably associated with the different muscle fiber compositions and different training programs of these athletes.

Comparison With Anaerobic Indices Measured in the Laboratory

Correlation coefficients between power indices of the WAnT and other variables measured in the laboratory are summarized in table 3.4. These variables

Table 3.2 Correlation Between the Wingate Anaerobic Test Scores and the Performance in "Anaerobic" Performance Tasks

WAnT index	N and gender	Exercise	r	Comments	Reference
PP	35M	40m run speed	0.84	10- to 15-year-olds, random sample	Bar-Or and Inbar (1978)
PP	9M	50m run time	-0.91	Active young adults	Kaczkowski et al. (1982)
PP	24M	4 3 91.5m skate	0.83	10-year-old ice hockey players	Rhodes et al. (1985)
PP · kg^{-1}	56M	50-yard run time	-0.69	10- to 15-year-olds, active	Tharp et al. (1985)
PP	56M	Vertical jump	0.70	10- to 15-year-olds, active	Tharp et al. (1985)
PP	87M	500m skate speed	0.66	US National team	Thompson et al. (1986)
PP	24M	SAS40	0.32	Junior A ice hockey players	Watson and Sargeant (1986)
MP	9F&M	25m swim time	-0.90	8- to 12-year-old swimmers	Inbar and Bar-Or (1977)
MP	9F&M	25m swim time	-0.92	8- to 12-year-olds, WAnT arm	Inbar and Bar-Or (1977)
MP	22F&M	300m run time	-0.88	8- to 12-year-old swimmers	Inbar and Bar-Or (1977)
MP	35M	300m run speed	0.85	10- to 15-year-olds, random sample	Bar-Or and Inbar (1978)
MP	24M	4 3 91.5m skate	0.71	10-year-old ice hockey players	Rhodes et al. (1985)
MP · kg^{-1}	56M	50-yard run time	0.69	10- to 15-year-olds, active	Tharp et al. (1985)
MP	56M	Vertical jump	0.74	10- to 15-year-olds, active	Tharp et al. (1985)
MP	24M	SAS40	0.79	Junior A ice hockey players	Watson and Sargeant (1986)
MP	87M	500m skate speed	0.76	US National team	Thompson et al. (1986)
MP	10M	300m cycle time	<-0.75	25.7-year-old cyclists	Perez et al. (1986)

Note: PP = peak power, MP = mean power, SAS40 = Sargeant Anaerobic Skating test

Reprinted from Bar-Or 1987.

Table 3.3   Comparison of Measures Obtained During the Wingate Anaerobic Test on Five Groups of Highly Trained Athletes; Mean ± (Standard Deviation)

Group	Peak power $(W \cdot kg^{-1})$	Mean power $(W \cdot kg^{-1})$	Fatigue index (%)
Power lifters	12.6 (1.0)	9.3 (1.3)	45 (8.5)
Gymnasts	12.3 (0.7)	9.1 (0.7)	47 (3.5)
Wrestlers	12.0 (0.9)	9.3 (0.9)	43 (5.2)
10-km runners	11.9 (0.6)[a]	9.3 (0.8)	33 (7.2)[a,b]
Ultramarathoners	11.2 (1.1)[a]	8.8 (0.6)	26 (8.7)[a,b]

[a]Different from power lifters ($p < 0.05$)
[b]Different from gymnasts and wrestlers ($p < 0.05$)

Table 3.4   Correlation Between the Wingate Anaerobic Test Scores and Other Laboratory Anaerobic Indices

WAnT index	N	Laboratory test or index	r	References
PP	15	Margaria step-running	0.79	Ayalon et al. (1974)
PP · kg⁻¹	11	Margaria step-running	0.84	Jacobs (1979)
PP · kg⁻¹	15	Margaria step-running	–0.003	Taunton et al. (1981)
PP	19	PP—Thorstensson isokinetic	0.61	Inbar, Kaiser, and Tesch (1981)
MP	19	MP—Thorstensson isokinetic	0.78	Inbar, Kaiser, and Tesch (1981)
MP	16	Maximal O_2 debt	0.86	Bar-Or et al. (1977)
PP	11	O_2 debt post WAnT	0.85	Jacobs (1979)
MP	11	O_2 debt post WAnT	0.63	Jacobs (1979)
MP	14	O_2 debt post WAnT	0.47	Tamayo et al. (1984)
rev/30 s	11	Lactate post WAnT	0.60	Jacobs (1979)
MP · kg⁻¹	14	Lactate post WAnT	0.60	Tamayo et al. (1984)
PP · LBM⁻¹	19	% FT area	0.60	Bar-Or et al. (1980)
% fatigue	19	FT area/ST area	0.75	Bar-Or et al. (1980)
MP · LBM⁻¹	19	FT area/ST area	0.63	Bar-Or et al. (1980)
PP	29	% FT	0.72	Inbar, Kaiser, and Tesch (1981)
MP	29	% FT	0.57	Inbar, Kaiser, and Tesch (1981)
PP	9	FT area	0.84	Kaczkowski et al. (1982)
MP	9	FT area	0.83	Kaczkowski et al. (1982)

Note: PP = peak power, MP = mean power, LBM = lean body mass; FT = fast-twitch muscle fiber, ST = slow-twitch muscle fiber
Reprinted from Bar-Or 1987.

include two performance tasks (the Margaria step-running test and the Thorstensson 60-s isokinetic knee-extension test), maximal oxygen debt, maximal blood lactate, and the distribution and cross-sectional area of muscle fiber types.

In two studies (Ayalon, Inbar, and Bar-Or 1974; Jacobs 1979), PP and the power measured by the Margaria step-running test seem closely associated. The lack of any relationship between these two indices in the study by Taunton, Maron, and Wilkinson (1981) is surprising, but that study also found extremely low correlations among several other anaerobic indices; this may reflect sample specificity.

The Thorstensson test includes 50 isokinetic all-out knee extensions in 60 s. While the correlation coefficients summarized in table 3.4 are significant, some of the common variance might be related to body mass. Nevertheless, the coefficients remained significant, even when performance in both tests was expressed per kg body mass.

Both maximal O_2 debt and maximal blood lactate have long been considered anaerobic indices. There appears to be only one study in which maximal O_2 debt measured following a treadmill test for $\dot{V}O_2$max was related to the WAnT (Bar-Or, Dotan, and Inbar 1977). The correlation coefficient with MP was 0.86. Other authors did not use maximal values of O_2 debt, only submaximal values elicited by the WAnT. Oxygen debt during recovery from the WAnT is only 80% of that obtained following a treadmill test of maximal aerobic power (Inbar, Dotan, and Bar-Or 1976; Jacobs 1979). The same distinction must be made when analyzing the reports by Jacobs (1979) and Tamayo et al. (1984), who correlated MP with submaximal blood lactate values.

Comparison With Histochemical Findings

Athletes who specialize in events requiring high mechanical power have a higher ratio of fast-twitch (FT) to slow-twitch (ST) muscle fibers than do endurance athletes. FT fibers typically generate higher mechanical tension and fatigue earlier than ST fibers (Burke, Levine, and Zajac 1971; Thorstensson and Karlsson 1976). If the WAnT indeed measures anaerobic capability, it is reasonable to assume that people with high scores will also have a high FT/ST ratio and a high FT-area/ST-area ratio. This hypothesis was first tested among 19 male Israeli physical education students, sprinters, and long-distance runners (Bar-Or et al. 1980). Expressed per kg lean body mass, PP correlated significantly ($r = 0.60$) with %FT area. Both MP and the fatigue slope correlated significantly ($r = 0.63$ and 0.76, respectively) with the ratio of FT area/ST area (figs. 3.2 and 3.3). Similar relationships were found among 29 young Swedish adult males who ranged in activity level from being sedentary to competitive runners (Inbar, Kaiser, and Tesch 1981). Even higher correlations ($r = 0.83$ and 0.84 between the relative area of FT fibers and MP and PP, respectively) were found among nine physically active male Canadians (Kaczkowski et al. 1982). These values are similar to or higher than those found when performance on the Thorstensson 30-s isokinetic test was related to fiber-type distribution (Inbar, Kaiser, and Tesch 1981).

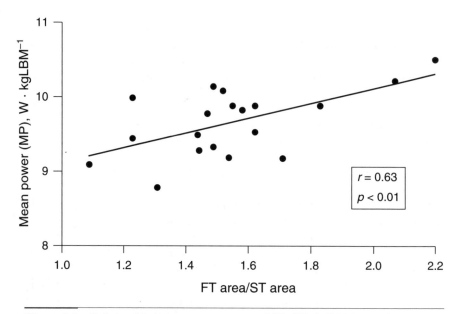

Figure 3.2   Relationship between mean power (MP) (W · kgLBM⁻¹) on the Wingate Anaerobic Test and relative fiber size (FT area/ST area).
Reprinted from Bar-Or et al. 1980.

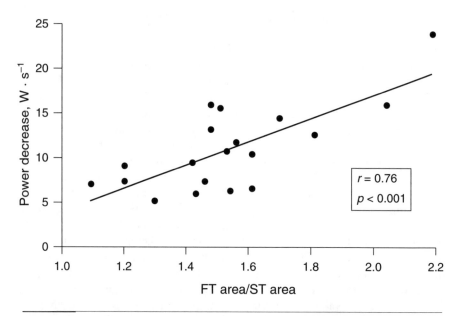

Figure 3.3   Relationship between power decrease (W · s⁻¹) in the Wingate Anaerobic Test and relative fiber size (FT area/ST area).
Reprinted from Bar-Or et al. 1980.

Histochemical analysis (Jacobs et al. 1983) showed a major surge in muscle lactate concentration (46.1 6 15.2 mmol · kg dry weight^{-1}) as early as 10 s from the start of the WAnT. At the end of 30 s, the value was 73.9 6 16.1 mmol · kg dry weight^{-1}. In another observation with female physical education students (Jacobs et al. 1982), the following concentrations (mmol · kg dry weight^{-1}) were found before and after the test, respectively: ATP, 20.9 and 13.8; CP, 62.7 and 25.1; lactate, 9.0 and 60.5; and glycogen, 360 and 278.

Denis et al. (1990) compared biochemical characteristics of the vastus lateralis muscle from highly trained sprinters and middle-distance runners. Significant relationships were found between WAnT performance and histochemical variables in the overall sample, with the highest coefficients found between PP and concentrations of lactate, hydrogen ions, and ATP in muscle after the WAnT (figs. 3.4, 3.5, and 3.6). Coupled with significant differences in the post-WAnT pH and lactate concentrations in the muscles of the two specialized trained athletes, these results point to the ability of the WAnT to detect specific anaerobic characteristics within relatively homogeneous populations specializing in a particular athletic event. This will be discussed further in chapter 5.

Based on these observations, the WAnT seems to tax the anaerobic energy pathways of both genders markedly. Performance of active athletic males in the WAnT seems to be related to the preponderance of their FT muscle fibers.

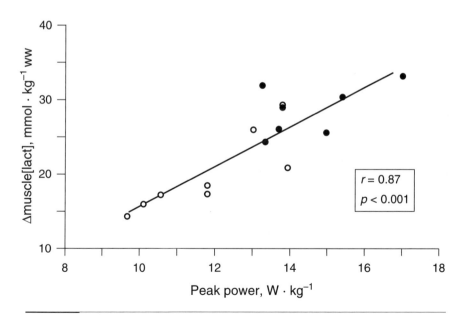

Figure 3.4   Relationship between changes (post-WAnT minus pre-WAnT) in muscle lactate (mmol · kg $^{-1}$ wet weight) and peak power output in 100-m sprinters (●) and 800-m middle-distance runners (○).
Reprinted from Denis et al. 1992.

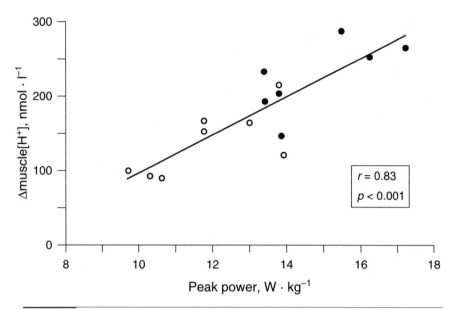

Figure 3.5   Relationship between changes (post-WAnT minus pre-WAnT) in muscle hydrogen-ion concentration [H⁺] and peak power output in 100-m sprinters (●) and 800-m middle-distance runners (○).
Reprinted from Denis et al. 1992.

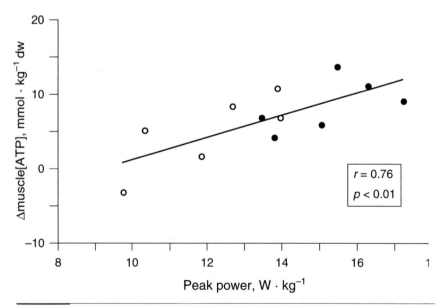

Figure 3.6   Relationship between changes (post-WAnT minus pre-WAnT) in muscle ATP concentration per kg dry weight and peak power output in 100-m sprinters (●) and 800-m middle-distance runners (○).
Reprinted from Denis et al. 1992.

Relative Aerobic and Anaerobic Contributions

It is wrong to assume that a certain task can be performed exclusively by either aerobic or anaerobic energy sources. This notion has been accepted for all-out protocols of maximal aerobic power that are also known to tax the anaerobic energy sources. Similarly, one can expect part of the energy for performing the WAnT to be derived from aerobic pathways. However, the anaerobic contribution must be predominant if the WAnT is to be considered "anaerobic."

To calculate the relative aerobic contribution, one should measure the net $\dot{V}O_2$ during the test and relate it to the total required biochemical energy. The latter can be calculated from the mechanical work produced during the test, assuming a certain mechanical efficiency. While mechanical efficiency during a submaximal aerobic task can be taken as 20-25%, there is no information on the actual mechanical efficiency of such a highly supramaximal task as the WAnT, in which mechanical power is two to four times that associated with the maximal aerobic power. This mechanical efficiency will probably be lower than that measured during submaximal tasks, however, and this is probably closer to 16-20%.

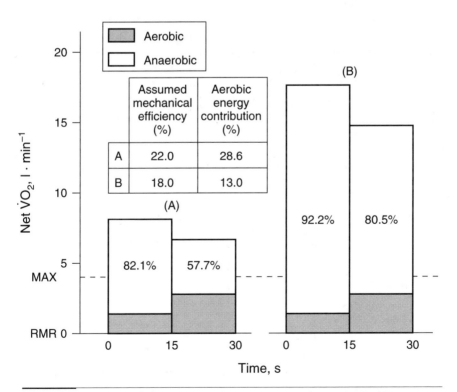

Figure 3.7 Aerobic and anaerobic energy contributions during the WAnT, assuming a mechanical efficiency of (a) 22% and (b) 18%.

Using a mean of both assumed mechanical efficiencies, Inbar, Dotan, and Bar-Or (1976) collected expired gas before, during, and after the WAnT in 16 males and females ages 16-22. Figure 3.7 is a summary of those findings. Assuming a mechanical efficiency of 22%, as in submaximal exercise (part A), the aerobic energy contribution in these subjects would be 28.6%. If one assumes a lower mechanical efficiency of 18% (part B), in which the O_2 deficit is considered equal to the measured O_2 debt, then the aerobic contribution would be only 13%. In all likelihood, the actual aerobic contribution in the WAnT is somewhere between these two values. Similar values can be derived from data reported by Jacobs (1979). A similarly low aerobic portion was calculated using data obtained during a 30-s supramaximal cycling test (McCartney, Heigenhauser, and Jones 1983).

Stevens and Wilson (1986) calculated the aerobic portion of the WAnT to be 44%. Personal communication with B.W. Wilson revealed a calculation error, however. When recalculated assuming a mechanical efficiency of 25%, the aerobic portion was 27%.

Another approach to assessing the relevance of aerobic metabolism to the WAnT was reported by Kavanagh et al. (1986). Their subjects performed the test while breathing room air and while breathing 12% O_2 (starting 5 min before the test). The latter regimen induced the arterial O_2 saturation to drop from 97.2% to 88.6%. As seen in figure 3.8, there was an insignificant change in MP and in PP, in spite of the hypoxic conditions. While this study could not quantify the aerobic contribution to the WAnT, it is apparent that the effect of hypoxia on performance was marginal.

Bedu et al. (1991) compared 47 Bolivian children acclimatized to high altitude (3,700 m) with 101 French children living at low altitude (330 m). The maximal power reached in 4-8 s using the Force-Velocity test of Pirnay and Crielaard (1979) was the same for the two groups, but there was a lower MP on the WAnT in the group at high altitude; this drop in MP was thought to be the result of a reduced glycolytic metabolism and a decreased aerobic energy

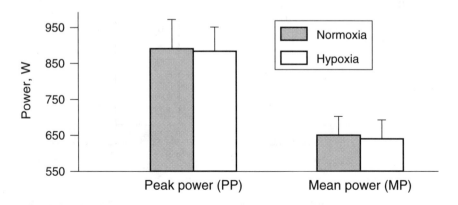

Figure 3.8 Peak power (PP) and mean power (MP) generated during the WAnT under normoxic (21% O_2) and hypoxic (12% O_2) conditions.

contribution attributed to the reduced $\dot{V}O_2$max seen at altitude. Unfortunately, the same subjects were not studied at both altitudes, so it is not possible to know if the reduced MP was caused by differences between the groups.

Sensitivity

If a test is to be informative, it must be sensitive to changes in the measured component. In the case of the WAnT, the component of interest is anaerobic performance and the changes that are expected as a result of such factors as health and activity. Although the effects of bed rest or inactivity on WAnT performance have not been examined, the effects of various diseases and physical training have been studied extensively.

In a study by Grodjinovsky et al. (1980), 45 boys ages 11-12 years participated in an 8-week mild conditioning program. Fifteen boys trained by doing short all-out pedaling bouts on the cycle ergometer, while another 15 boys ran 40-m and 150-m sprints. A third group did not train and served as controls (fig. 3.9). MP and PP of the training groups improved significantly (3-5%), and the control group did not change. In another study, Inbar and Bar-Or (1980) trained young adults intensively for 8 weeks on the cycle ergometer

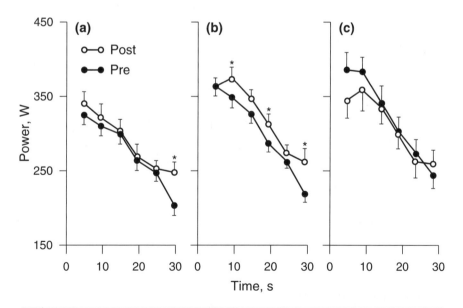

Figure 3.9   Power generation ($\bar{X} \pm SD$) during WAnT before (●) and after (○) 8 weeks of (a) anaerobic-type cycling or (b) sprinting compared with (c) no training (*$p <$ 0.05).

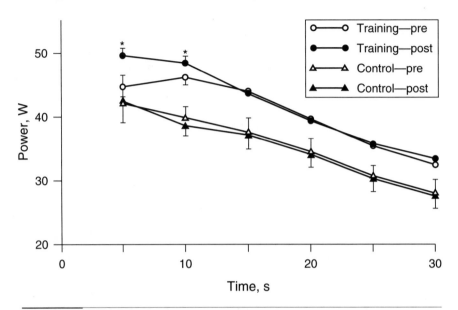

Figure 3.10 Effect of anaerobic-type leg training on power development ($\bar{X} \pm SD$) during the WAnT with legs (*$p < 0.05$) compared with results from untrained control subjects.
Reprinted from Inbar and Bar-Or 1980.

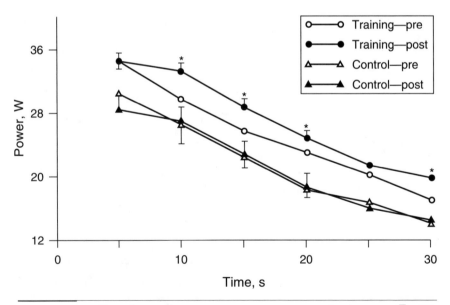

Figure 3.11 Effect of anaerobic-type arm training on power development ($\bar{X} \pm SD$) during the WAnT with arms (*$p < 0.05$) compared with results from untrained control subjects.
Reprinted from Inbar and Bar-Or 1980.

using either the arms or the legs. Similar increases in PP and MP were found in the specific muscle groups trained (figs. 3.10 and 3.11). Somewhat greater improvement in PP and MP (8-10%) was achieved by girls and boys 12-13 years old who took additional physical education classes at school during an 8-month period (Grodjinovsky and Bar-Or 1984).

Denis et al. (1990) attempted to evaluate the effects of sprint and endurance training on WAnT performance. Figure 3.12 summarizes their results and confirms the sensitivity of WAnT to detect overall changes resulting from training, as well as those changes specific to the different types of training. Endurance training had no effect, whereas sprint training significantly improved anaerobic performance during the first 6 s.

It should be mentioned that the improvement in the WAnT indices following such short-term training regimens (3-15%, depending on the initial levels and on the training intensity) is appreciably lower than that known to occur in $\dot{V}O_2$max following aerobic training programs of similar length (Bouchard et al. 1988).

No studies have systematically documented the sensitivity of the WAnT to detraining. Anecdotal clinical data, however, suggest that deterioration in a subject's physical function is accompanied by a reduction of scores in the WAnT. A case in point is a boy with progressive muscular dystrophy who was followed periodically at the Children's Exercise and Nutrition Centre (Bar-Or 1993). When the child maintained an active lifestyle over a period of several months,

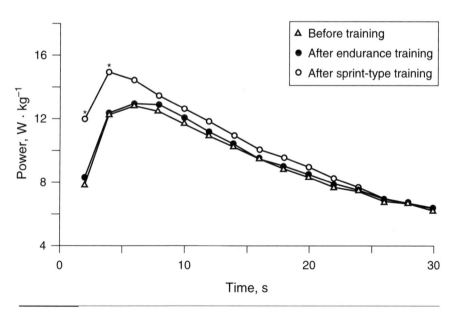

Figure 3.12   Effect of endurance (●) and sprint-type (○) training on power development during the WAnT with legs (*$p < 0.05$). Results before training (△) are shown for comparison.
Reprinted from Denis et al. 1990.

his anaerobic performance was quite stable. However, after 3-4 weeks of bed rest following surgery, he could no longer walk and his peak PP and MP decreased by 50% to 60%.

Thus, it seems that the WAnT is sensitive and mode-specific to increases in anaerobic performance achieved by various physical training modes. Further information is still needed to ascertain whether the test is sensitive to the short-term and long-term effects of detraining. Such information would be important for rehabilitation and for coaching.

In conclusion, even when performed under nonstandardized, warm, or humid climatic conditions, as well as during mild to moderate hypohydration, the WAnT is reliable and reproducible. Motivational factors involving emotional arousal may increase PP but not MP. Conversely, a 15-min warm-up may increase MP but not PP. Although the WAnT is highly reliable and reproducible, preparation for the test and its execution should be carefully standardized.

Chapter 4

Factors That May Influence Performance of the Wingate Anaerobic Test

As discussed in chapter 3, the WAnT is a reliable, valid, and sensitive test of anaerobic performance under controlled, standardized conditions. Because persons using the WAnT need to know which factors should be controlled or at least considered, this chapter reviews the effects of warm-up, climate, circadian rhythm, hypohydration, motivation, acid-base status, and physical training on various aspects of WAnT performance.

Warm-Up

Although warming up is recommended before engaging in strenuous activity, only the study by Inbar and Bar-Or (1975) has systematically assessed the effects of warming up on performance of the WAnT (fig. 4.1). Boys 7-9 years old who were naive to the concept of warm-up performed the test with and without a 15-min intermittent (30 s on, 30 s off) warm-up on the treadmill. Warming up consistently and significantly improved MP by 7%, but this did not affect PP. Unpublished data on these subjects suggested that the intermittent warm-up was more effective than a 15-min equicaloric continuous warm-up. More data are undoubtedly needed to construct an optimal warming-up protocol for the WAnT. This may depend on such factors as the age, gender, and fitness of the subjects, their state of health, and the prevailing climate.

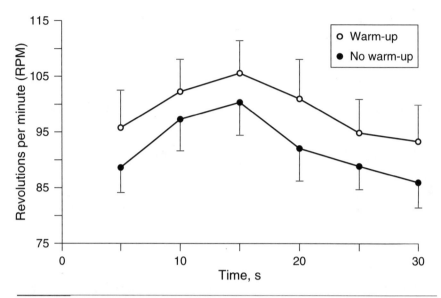

Figure 4.1 Changes in rate of pedaling (RPM) during WAnT with (○) and without (●) warm-up ($\bar{X} \pm SD$).
Reprinted from Inbar and Bar-Or 1975.

Climate

As the WAnT is designed to be used both in the laboratory and under field conditions, it is pertinent to ask whether climatic conditions must be standardized for its administration, especially since Bergh (1980) found a 5% drop in anaerobic performance for each 1°C drop in core temperature. Comparisons have been made of performances of 28 children, ages 10 to 12, who were tested in thermoneutral (22° to 23°C, 55% to 60% relative humidity or R.H.), hot-dry (38° to 39°C, 25% to 30% R.H.), and warm-humid (30°C, 85% to 90% R.H.) environments (Dotan and Bar-Or 1980). All tests were performed following a 45-min exposure to the respective climate. Test-retest correlation coefficients for MP ranged from 0.89 to 0.93 (figs. 4.2 and 4.3). Means for PP ($W \cdot kg^{-1}$) were 7.87 for the thermoneutral condition, 7.96 for the warm-humid climate, and 7.87 for the hot-dry climate. Respective values for MP were 6.82, 6.92, and 6.74 $W \cdot kg^{-1}$. These interenvironmental differences were insignificant. Thus, it seems that exposure to hot-dry and warm-humid climates does not affect children's performance. Data are still needed on other age and gender groups, as well as on the possible effect of cold climates.

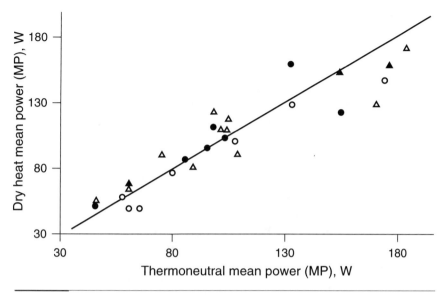

Figure 4.2   Mean power (MP) output during the WAnT when performed under thermoneutral (22° to 23°C; 55% to 60% R.H.) and dry heat (38° to 39°C, 25% to 30% R.H.) climatic conditions. The line drawn is the line of identity. Boys: △ = warm-up; ▲ = no warm-up; Girls: ○ = warm-up; ● = no warm-up.

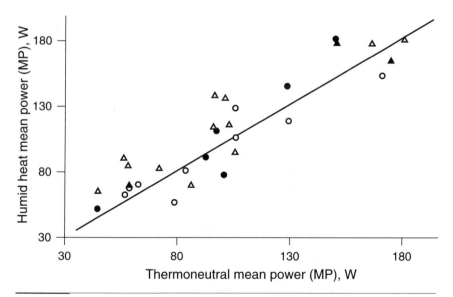

Figure 4.3   Mean power (MP) output during the WAnT when performed under thermoneutral (22° to 23°C; 55% to 60% R.H.) and humid heat (30°C; 85% to 90% R.H.) climatic conditions. The line drawn is the line of identity. Boys: △ = warm-up; ▲ = no warm-up; Girls: ○ = warm-up; ● = no warm-up.

Circadian Rhythm

Circadian rhythms refer to cyclical changes over approximately 24 h. Such rhythms are found in many biological functions, e.g., blood pressure (Cabri et al. 1988), and are closely associated with the circadian rhythm of body temperature (Reilly 1987), which is lowest in the early morning and highest in the late afternoon and early evening. Research has shown that exercise performance varies with time of day (Baxter and Reilly 1983; Reilly 1987). For example, Reilly and Baxter (1983) found that when a large muscle mass was active and a fixed high-intensity exercise had to be sustained for as long as possible, the performance in the evening was superior compared to that in the morning. An explanation for such findings may be found in the diurnal variation in body temperature, as Bergh (1980) showed that maximal anaerobic exercise was impaired by about 5% for each 1°C fall in core temperature. Thus, performance on the WAnT might vary with the time of the day.

One unpublished study by Wilby (personal communication), however, found no meaningful circadian rhythm–related changes in the WAnT indices when tested in duplicate every 4 h for 24 h. These findings are in agreement with results reported for arm ergometry over 30 s (Reilly and Down 1986) and for eccentric and concentric isokinetic leg exercises using a computer-linked dynamometer (Reilly 1987) and suggest that all-out dynamic ergometric performance, such as that required by WAnT, is independent of the time of day. Thus, more research appears warranted. In the meantime, it would appear prudent to do repeat tests with the WAnT at the same time of day.

Hypohydration

People who have been exercising and/or perspiring lose body fluids and may go from a state of euhydration (normal levels of body fluids) to a state of hypohydration (low levels). Given that an adequate fluid balance is important for the proper functioning of the body, there is a need to know whether fluid balance affects the performance of brief, high-intensity exercise.

Eleven members of a university wrestling team performed the WAnT in the euhydrated state and at three levels of hypohydration corresponding to 2%, 4%, and 5% of initial body weight (Jacobs 1980). Passive thermal dehydration was induced on separate days by exposure to 56°C, 10% to 20% R.H., following which subjects rested in a neutral environment for 30 min and then performed the WAnT. Average PP with no hypohydration was 859 W, compared with 840, 841, and 839 in the 2%, 4%, and 5% conditions, respectively. The respective values for MP were 639, 644, 631, and 636 W. None of these differences were significant, nor were there any differences in postexercise blood lactate. It should be stated here that these conclusions cannot be generalized directly to other groups nor to higher degrees of hypohydration.

Motivation

Motivation may play a role in the performance of any all-out task. It was therefore important to determine whether performance on the WAnT was affected by environmental manipulations that may modify motivation. In a study by Geron and Inbar (1980), seven types of motivation were given to young adult nonathletes: (1) presence of an audience, (2) individual competition, (3) group competition, (4) punishment, (5) reward, (6) group association, and (7) social responsibility.

The main finding was that motivational stimuli based on cognitive information had little or no effect on WAnT performance. In contrast, motivation based on such emotional factors as reward and punishment may improve performance, particularly of PP. Experience has shown that conventional encouragement and feedback during the test may not affect its outcome. It is of immense importance, however, to get a subject's cooperation by explaining the nature and importance of the test. Conclusions based on the study by Geron and Inbar (1980) may be limited to active, nonathletic young adults, however. Until further research is done with other groups, environmental conditions that may affect motivation should be consistent.

Changes in Acid-Base Status

Accumulation of intramuscular hydrogen ions (H^+) seems to be a major factor in fatigue during short-term, high-intensity exercise (see a review by Parkhouse and McKenzie 1984), but less so in prolonged, aerobic exercise. Among its other effects, acidosis interferes with the rate of anaerobic glycogenolysis.

Because high-intensity exercise increases the concentrations of lactate (LA) and H^+ in blood and muscle, and because these high levels of acidity adversely affect energy conversion, there have been numerous attempts to reduce acidity with buffers, especially sodium bicarbonate. This compound is used because it is thought to delay fatigue by increasing the gradient and diffusion rate of LA and H^+ from muscle to blood (Mainwood and Cechetto 1980).

In a cooperative study between groups from the Wingate Institute and the Karolinska Hospital in Stockholm (Inbar et al. 1983b), 13 male physical education students performed the WAnT twice. On one occasion, they ingested 10 gm of sodium bicarbonate ($NaHCO_3$) in capsule form 3 h before the test. On another day, they received a placebo. Blood pH just before the test and also following exercise was significantly higher with $NaHCO_3$ (7.43 vs. 7.37 before exercise and 7.23 vs. 7.18 after exercise). Although alkalinization did not affect PP, it did induce a small but significant increase in MP (fig. 4.4). Similar results were obtained by Sharp et al. (1986), who administered 30 gm · kg body mass^{-1} of $NaHCO_3$ to eight males 3 h before a modified WAnT lasting 40 s. Although PP was not affected, MP was higher with $NaHCO_3$.

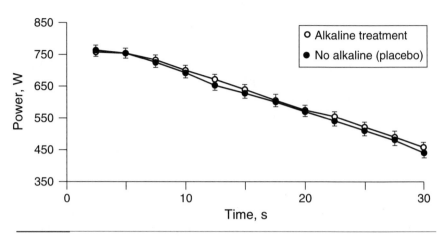

Figure 4.4   Power output every 2.5 s during WAnT with ($\bigcirc$) and without ($\bullet$) alkaline treatment before the test. Data points are $\bar{X} \pm SEM$.
Reprinted from Inbar et al. 1983.

The same subjects also performed a longer task (cycling to exhaustion at an intensity equivalent to 150% $\dot{V}O_2$max), but alkalinization did not affect time to exhaustion. Acidosis caused by NH_4CO_3 ingested 3 h before the test did not affect performance on either test.

Horswill et al. (1988) found that the total work output in 2 min did not differ when subjects were given 0, 10, 15, and 20 gm $NaHCO_3$ per kg body mass, even though plasma HCO_3^- levels were higher and related to the dosage given. Reviewing other studies, Horswill et al. (1988) concluded that bicarbonate loading might be effective if higher dosages (e.g., > 30 gm · kg^{-1}) and/or repeated bouts of near-maximal exercise were studied. Nevertheless, many subjects have gastrointestinal distress and diarrhea within 1 h of ingestion, which suggests that taking bicarbonate is not a practical way to improve performance of this type of exercise.

Physical Training

Based on the established sensitivity of the WAnT to detect changes brought about by training, this test has been used in many studies that address the problem of training effects on anaerobic performance. A brief review of some of these studies is presented below.

Many types of programs have been studied to determine the effects of training on anaerobic performance. Most of these have used interval running or cycling at maximal or near-maximal intensities, but there were large differences in the number of repetitions, the duration of the work and rest periods, and the ratio of work to rest. Many different tests of anaerobic performance have also been used. As a result, evaluating the effects of training on anaerobic performance is somewhat complex, especially as it relates to the WAnT. If the evaluation is based primarily on the intensity and duration characteristics of the WAnT, however, the task is simplified.

Given that the PP is associated with very high-intensity anaerobic exercise for a few seconds and that the MP is associated with high-intensity anaerobic exercise for less than 60 s, a pattern can be seen. Table 4.1 summarizes seven training programs and their results. None used the WAnT as the criterion measure of anaerobic performance.

Looking at the first two studies in table 4.1, in which the duration of the intermittent training bouts was similar to that of the PP on the WAnT (5 s and 10-12 s, respectively), both reported significant improvements in leg strength, height jumped, and myokinase activity, i.e., factors associated with brief, high-intensity exercise. Following sprint training, Thorstensson, Sjödin, and Carlsson (1975) also found increases in creatine phosphokinase (CPK) and ATPase activities, but a later study by Thorstensson et al. (1976) found no change in these phosphagen enzymes after strength training. Although the

Table 4.1   Studies on Training and Anaerobic Performance

Description of training program	Duration of training bouts	Effect on factors related to PP	MP	References
Run 20-40 reps, 19-24 km/h, 9-10% grade	5 s	+	0	Thorstensson et al., 1975
3 sets of 6 maximal lifts	10-12 s	+	0	Thorstensson et al., 1976
10-15 rides, 1.5-2 · $\dot{V}O_2$max (60 or 120 rpm)	20 s	+	+	Campbell et al., 1979
2 all-out rides at 4 kg	40 s	+	+	Weltman et al., 1978
8 200-m runs at 90% max	25-30 s		+	Roberts et al., 1982
≤ 19 runs at 15-17 km/h, 5-10% grade	30 s	0		Fox et al., 1977
≤ 7 runs at 10-12 km/h, 5-12% grade	120 s			
5 sprints, 44% grade	6 s	+/0	+	Houston and Thomson, 1977
15 maximal leg presses	30 s			
3 maximal runs, 3.3% grade	60 s			
2 maximal runs, 3.3% grade	90 s			

Note: Blank = not measured; 0 = no change; + = improvement
Reprinted from Skinner and Morgan 1985.

sprint training study found no change in ATP or phosphocreatine (CP) concentrations, they did find a rise in the total amount of these phosphagens due to a larger muscle mass. Possibly because the training bouts were so short, no difference was found in lactate dehydrogenase (LDH) or phosphofructokinase (PFK), two enzymes involved in anaerobic glycolysis, which occurs in exercises lasting longer than 10 s.

Following intermittent 20-s training bouts on the cycle ergometer (a duration between that of PP and MP on the WAnT), Campbell et al. (1979) reported significant increases in the Sargent jump and in the 20-s power output on the cycle and found improvements in both peak 4-s and total 40-s power outputs. After subjects ran 25-30 s, Roberts, Billeter, and Howald (1982) reported an increase in treadmill run to exhaustion (47-58 s) and increased activity of the following glycolytic enzymes: PFK, phosphorylase, GAPDH, LDH, and MDH. The lack of change in succinate dehydrogenase (SDH) activity suggests that the mitochondrial oxidative enzymes were not affected.

Houston and Thomson (1977) had endurance-adapted subjects train at intervals of 6, 30, 60, and 90 s. Significant increases were noted for leg strength (35%) and ATP concentration in the muscle (15%), but there were no differences in CP concentration or in the performance of the Margaria test. Treadmill run time to exhaustion improved 17% (from 47 s to 55 s), and there was a significant rise in the average distance during the 60-s (14%) and 90-s (13%) runs.

In an attempt to clarify the specificity and generality of training, five training studies were done by O'Connor (1987) using different combinations of exercise intensities and durations designed to stimulate the anaerobic and/or aerobic systems. Table 4.2 outlines the training regimens that ranged from very brief anaerobic exercise to moderate aerobic exercise.

As expected, no changes occurred in $\dot{V}O_2$max or in performance on the WAnT by the control group over 8 weeks (see table 4.3). Results from the WAnT showed that the 10-s maximal anaerobic training produced a significant rise in PP and MP, the 30-s very high anaerobic training caused a significant increase in MP, and the 2-min mixed program improved both PP and MP.

Table 4.2   Five 8-Week Cycle Ergometer Training Programs Three Times a Week

Type of training	Intensity	Total duration per session, (work:recovery)
Maximal anaerobic	80% PP (WAnT)	40 min (10:30 s)
Very high anaerobic	80% MP (WAnT)	40 min (30:90 s)
Aerobic and anaerobic	120% $\dot{V}O_2$max	40 min (2:6 min)
High aerobic	80% $\dot{V}O_2$max	30 min
Moderate aerobic	60% $\dot{V}O_2$max	60 min

Table 4.3   Results of Five 8-Week Cycle Ergometer Training Programs

Program	PP	WAnT (W · kg^{-1}) MP	$\dot{V}O_2$max (ml · kg^{-1} · min^{-1})
Control (8 males)			
Before	10.7	8.2	43.2
After	10.6	7.9	41.6
Maximal anaerobic (9 males)			
Before	10.9	7.8	41.8
After	12.1[a]	9.0[b]	44.4[b]
Very high anaerobic (6 males)			
Before	11.2	8.1	42.9
After	12.0[c]	9.2[b]	48.9[a]
Aerobic and anaerobic (9 males)			
Before	11.4	7.7	46.0
After	12.6[a]	9.2[a]	52.4[a]
High aerobic (8 females)			
Before	9.4	6.7	36.0
After	9.6	7.2	42.4[a]
Moderate aerobic (8 males)			
Before	10.5	7.5	37.9
After	11.0	8.0	43.4[a]

[a]$p < 0.01$; [b]$p < 0.05$; [c]$p < 0.07$

The two aerobic training programs had little effect on either PP or MP. On the other hand, all five programs produced a significant rise in aerobic ability (as reflected by $\dot{V}O_2$max), but the lowest effect was seen in the 10-s program of maximal anaerobic training. The rise in $\dot{V}O_2$max with the anaerobic programs probably results from the fact that the rest intervals were too short to allow the aerobic system to return to baseline before the next exercise interval began. Thus, it appears that continuous aerobic training has a specific effect on aerobic ability, whereas intermittent anaerobic training has a more general effect on both anaerobic and aerobic abilities.

Studies indicate that heredity plays an important role in the response to training. There are large individual differences in the ability to train one's aerobic and anaerobic abilities. For example, while mean training gains in aerobic ability ($\dot{V}O_2$max and maximal work output in 90 min) were 33% and 51%, respectively, the variation in gains ranged from 10% to 90%. Anaerobic ability (maximal work outputs in 10 and 90 s) improved an average of 22% and 35%, respectively, ranging from 5% to 70% (Bouchard et al. 1988). Data from many studies also suggest that there are low responders and high responders, as well as those who respond early and those who respond late to training. The size of the genetic effect is about 40% for $\dot{V}O_2$max (ml · kg^{-1} · min^{-1}) and 60% to 70% for the 90-min total work output, i.e., the effect is greater for endurance performance than for maximal aerobic power (Bouchard et al. 1988). The difference in response to endurance training is 65% to 80% genotype dependent, especially in the latter stages of long-term training when subjects approach their genetic limits of adaptation (Hamel et al. 1986). Simoneau et al. (1986) looked at the genetic effect on training responses to brief anaerobic exercise (10-s work output) and to longer anaerobic exercise (90-s work output). The genetic effect was low (30%) for the 10-s exercise but significant (70%) for the 90-s bout.

Summarizing these studies, it appears that improvements in activities related to PP are primarily associated with increases in the amount of ATP and its rate of degradation. Improvements in those activities related to MP are essentially brought about by increasing the rate of glycolysis, resulting in increases in power output, maximal lactate concentrations, and maximal oxygen debt for activities lasting 20-45 s. Although some of these results may appear ambiguous and equivocal, the type of adaptation tends to be related to the duration of the training bouts, i.e., the effects are specific to the type(s) of anaerobic ability being stimulated. Unfortunately, test durations were usually similar to those of the training bouts. As a result, there may have been changes in other types of anaerobic performance, but these were often not measured.

Chapter 5

Typical Values of the Wingate Anaerobic Test

To help the reader understand and interpret the WAnT, it is useful to show some typical values found with various groups of subjects. We have purposely chosen not to call these "norms" or "reference standards" because they often have been collected on relatively small, select, cross sections of the population that may not represent the general population from which they came. This chapter presents typical values found with different age groups, with both sexes, with different sports, with persons suffering a chronic disease, and with other special populations.

Age

Abundant cross-sectional and longitudinal data are available on the effects of growth, development, maturation, and aging on maximal aerobic power. In contrast, only sparse information has been published on the age-related changes in one's ability to generate and sustain a high-intensity, mostly supramaximal power output for 30-45 s. So far, only the Margaria test and the WAnT have been used to study the anaerobic characteristics of relatively large populations over a wide age range.

Ideally, a study of the effects on physical performance of age, growth, and maturation should be based on a longitudinal study that encompasses prepubescent, pubescent, and postpubescent periods (Kemper et al. 1983; Rutenfranz 1986). There are, however, no such data regarding anaerobic performance and/or capacity.

The first report on age-related peak mechanical power was published by Margaria, Aghemo, and Rovelli (1966). They tested 131 healthy females and males, ages 8 to 73, using the Margaria step-running test. Of these subjects, 35 were athletes and 96 were sedentary. Children reached a distinctly lower peak power than the adolescents and young adults; this difference was apparent whether expressed in absolute power units or relative to body mass. Among the nonathletes, peak power per kg body mass in 9-year-olds was only 60% of that found in 20-year-olds.

Using the same test, Di Prampero and Cerretelli (1969) determined the peak power of 156 African (Nilo-Hamitic and Bantu) females and males, ages 5 to 68. Again, the children performed considerably worse than the adolescents or young adults, both in absolute power units and relative to body mass. Similar findings were reported by Davies, Barnes, and Godfrey (1972) on 92 British children, ages 6 to 16, and by Kurowsky (1977) on 294 American children, ages 9 to 15.

Cross-sectional data have also been gathered at the Wingate Institute (Inbar and Bar-Or 1986) using the WAnT on 306 active but untrained males, ages 8 to 45, with no apparent disease. After being subdivided into eight age groups, some performed both arm and leg tests, while others performed one or the other. Figure 5.1 summarizes performance on the WAnT among the males, as related to chronological age. It is apparent that PP and MP of both the legs and arms increased consistently from age 10 to young adulthood. The values seem to peak at the end of the third decade for the legs and at the end of the second decade for the arms. During the second decade, PP for the legs rises about 35 W · yr^{-1} and MP increases about 30 W · yr^{-1}. The corresponding increases for the arms are 32 and 22 W · yr^{-1}. As was shown in the Margaria step-running test (Margaria, Aghemo, and Rovelli 1966; Di Prampero and Cerretelli 1969), even when corrected for body mass, both PP and MP were lowest in the youngest age group and rose with age through adulthood. For example, the relative power output of a 10-year-old boy was 80% and 75%, respectively, of that generated by boys ages 13 and 17 years (see fig. 5.2).

As shown in figure 5.1, the anaerobic performance by the arms is 60% to 70% of that achieved by the legs at all ages. This percentage is similar to that described for young adults who did a maximal aerobic power test with their arms and with their legs, and it probably reflects the different muscle mass used in arm cranking and leg pedaling (Bar-Or & Zwiren 1975).

Numerous studies of males (Åstrand 1952; Krahenbuhl, Skinner, and Kohrt 1985; Robinson 1938; Shephard et al. 1969) have shown that $\dot{V}O_2$max (ml · kg^{-1} · min^{-1} is virtually independent of age between the ages of 8 to 30 years. Among females, it is even higher in the prepubescent phase than during or after puberty.

Unlike $\dot{V}O_2$max, anaerobic performance is lower in children, not only in absolute power units but also when corrected for body mass. A graphic comparison of the growth-related differences between maximal aerobic performance and PP or MP is shown in figure 5.2. To put aerobic and anaerobic performance on the same scale, values are presented in percentage units, taking the highest value during young adulthood as 100%.

In figure 5.2(a), MP and MP · kg^{-1} are plotted against $\dot{V}O_2$max (ml · kg^{-1} · min^{-1}) values taken from the literature (Bar-Or 1983; Robinson 1938). While $\dot{V}O_2$max remains virtually unchanged throughout the second and third de-

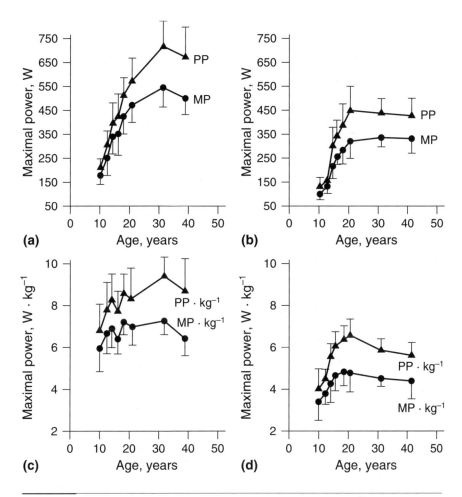

Figure 5.1   Anaerobic performance and age. Cross-sectional data ($\bar{X} \pm SD$) for absolute and relative power outputs (PP = 5-s peak power and MP = 30-s mean power) on 306 males who performed the WAnT with the legs and arms. (a) Absolute power (W) for legs; (b) absolute power (W) for arms; (c) relative power (W · kg^{-1}) for legs; and (d) relative power (W · kg^{-1}) for arms.
Reprinted from Inbar and Bar-Or 1986.

cades, MP · kg^{-1} at age 10 is only 85% of that achieved in young adulthood. The corresponding value for the arms is less than 70% (fig. 5.2b). PP · kg^{-1} for the legs at age 10 is also about 70% of its value at the end of the third decade, and that for the arms is only 60%. The absolute MP leg performance of a 10-year-old boy is 30% of that found in a young adult and less than 30% when performed by the arms. Similar percentage values are seen for PP.

In a recent study, Falk and Bar-Or (1993) periodically tested 36 pre-, mid-, and late pubescent boys over a period of 18-24 months. While the peak

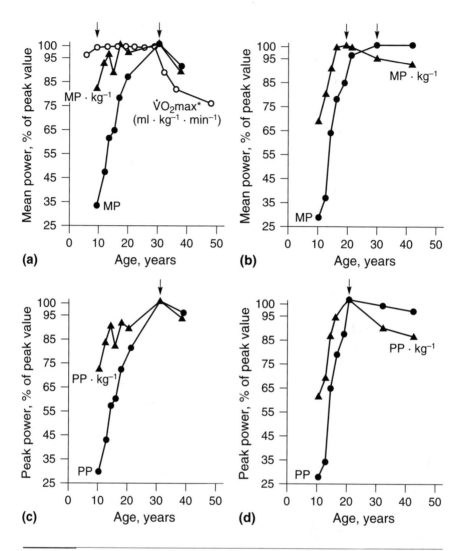

Figure 5.2 Anaerobic performance and age. Data are presented in percentages, taking the highest value of the adults as 100% (marked by arrows). Subjects and methods are the same as those in figure 5.1. (a) Absolute mean power (MP) and relative mean power (MP · kg⁻¹) for the legs; (b) same values for the arms; (c) absolute peak power (PP) and relative peak power (PP · kg⁻¹) for the legs; and (d) same values for the arms. (a) includes data from the literature (Robinson 1938) on maximal O₂ uptake (VO₂max)(○), also presented as a percentage of the adult value.
Reprinted from Inbar and Bar-Or 1986.

mechanical aerobic power (W · kg⁻¹) did not change with age or maturation, PP and MP per kg body mass increased over time. This pattern was particularly apparent during the transition from prepuberty to midpuberty.

Maximal aerobic power of children and adolescents, at least when measured during cycling, seems to be strongly related to the muscle mass that performs the activity (Davies, Barnes, and Godfrey 1972). Leg volume explains some variances in MP and PP among young adults (Murphy, Patton, and Frederick 1984).

Using the WAnT, MP and PP were measured during arm cranking in 50 girls and 50 boys ages 14 to 19 (Blimkie et al. 1988). As seen in figures 5.3 and 5.4, both MP and PP correlated more highly with lean arm volume (determined by water displacement and corrected for subcutaneous fat) and total body fat-free mass (estimated by skinfolds) than with chronological age or total body mass. In the boys, PP (and to a lesser extent, MP) increased progressively with age, even when expressed per unit of lean arm volume. Among the girls, on the other hand, there was no age-related increase in MP or in PP per unit of lean arm volume. Somewhat similar findings were described earlier by Di Prampero and Cerretelli (1969). In their study, the power generated

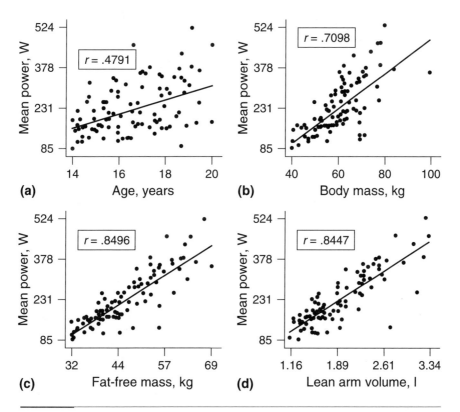

Figure 5.3 Relationship between mean power and (a) age, (b) body mass, (c) fat-free mass, and (d) lean arm volume. Individual data of 100 girls and boys who performed the arm WAnT.

Reprinted from Blimkie et al. 1988.

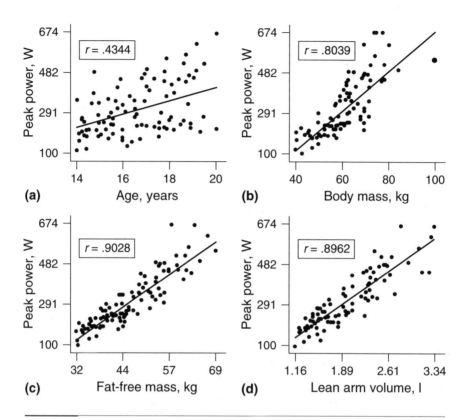

Figure 5.4   Relationship between peak power and (a) age, (b) body mass, (c) fat-free mass, and (d) lean arm volume. Individual data of 100 girls and boys who performed the arm WAnT.
Reprinted from Blimkie et al. 1988.

during the Margaria step-running test increased from childhood to adulthood in both males and females, even when corrected for fat-free mass. More recently, Sargeant (1989) has shown that 13-year-old boys generate less peak power per unit of lean thigh tissue than do young adults during a 30-s isokinetic cycling test.

All these observations suggest that age-related differences in anaerobic performance cannot be explained merely by differences in body size or active muscle mass. If this is the case, then age-related differences are probably best explained by qualitative characteristics of the muscle or by the nature of motor-unit activation. Some preliminary data on children's muscles suggest that their biochemical characteristics are different from young adults' muscles. Such differences are listed in table 5.1.

The concentration of ATP in resting muscle and its utilization during intense exercise seem to be similar in preadolescent boys and older males. By contrast, CP concentration is somewhat lower at rest in preadolescent boys

Table 5.1   Substrate Availability and Utilization in Muscles of Preadolescent Boys

	Resting values	Utilization rate during exercise	
	Concentration in muscle (mmol · kg^{-1} wet mass)	Compared with older individuals	
ATP	3.5-5	No change with age	Same as adults
CP	12-22	Lower in children	Same or less than adults
Glycogen	45-75	Lower in children	Much less than adults

Reprinted from Bar-Or 1983.

and has the same or a slightly lower rate of utilization than that found in older males. The main age- or maturation-related difference is in the concentration and rate of utilization of muscle glycogen. Both factors are distinctly lower in preadolescent boys (Eriksson 1980; Eriksson and Saltin 1974; Karlsson 1971). Based on these data, it seems that the biochemical difference in anaerobic characteristics between children and adults is associated more with anaerobic glycolysis and less with the phosphagen system.

This notion is supported by data on maximal muscle and blood lactate concentrations. In one study (Eriksson, Karlsson, and Saltin 1971), muscle lactate concentration immediately following maximal exercise was about 17 mmol · kg^{-1} in young adults and only 11 mmol · kg^{-1} in boys ages 13.5 to 14.8 years. A related finding is the lower concentration of the enzyme PFK in muscle of boys ages 11 to 13 (Eriksson, Gollnick, and Saltin 1973) and 16 to 17 (Fournier et al. 1982) compared to that seen in young adults. PFK is considered a rate-limiting enzyme in anaerobic glycolysis. Lower maximal blood lactate concentrations in children compared with adolescents or young adults has been found by many authors (Åstrand 1952; Krahenbuhl, Skinner, and Kohrt 1985; Matejkova, Koprivova, and Placheta 1980; Morse, Schultz, and Cassels 1949; Robinson 1938; Wirth et al. 1978) but not all (Cumming et al. 1980).

There are no data to explain why prepubescent children have a lower glycolytic capacity. It has been suggested that the rate of lactate production in rats (Krotkiewski, Kral, and Karlsson 1980) may depend on the level of circulating testosterone. One way to confirm this in humans is to correlate the rate of glycolysis with the level of testosterone in children of similar chronological age and body size but of different biological ages.

Another biochemical variable suggesting children's lower ability to perform anaerobic exercise is the maximal level of acidosis that they can reach. Whether expressed in pH units (Kindermann, Huber, and Keul 1975; Matejkova, Koprivova, and Placheta 1980; Von Ditter et al. 1977) or in base excess units (Matejkova, Koprivova, and Placheta 1980; Von Ditter et al. 1977), the level of acidosis during maximal effort is related to age. Between the ages of 8 and 18, there seems to be a decline of 0.01 to 0.02 pH units per year. The corresponding yearly decline in base excess is 1.0 to 1.5 mEq · L^{-1}. Lower values reflect higher levels of acidosis.

In conclusion, evidence is available on the reduced anaerobic performance of children; this poor performance is apparent during all-out brief step-running (up to 1 s) and during 30 s of cycling or arm cranking at supramaximal loads. Research is still needed on the specific developmental stage at which an individual acquires the adult characteristics for anaerobic exercise. More information is also needed on the anaerobic characteristics of females and on the trainability of anaerobic performance at different ages.

Although the biochemical data mentioned above are in line with the lower ability of children to perform anaerobic exercise, they do not explain the mechanism of such a deficiency. Studies at the cellular level are needed to tell whether there are any age- or maturation-related differences in muscle fiber types that are recruited during supramaximal exercise. However, such an approach requires invasive techniques that cannot be ethically justified in many countries. This constraint may limit the progress of our understanding of children's low anaerobic ability.

Age-related values of the Fatigue Index (FI) are not reported here because they followed a similar pattern to the relative peak power (PP · kg^{-1}) of any age group. It appears that the age-associated increase in FI is mainly the result of a parallel rise in PP, not of an aging phenomenon per se.

Gender

Such neuromuscular aspects as power, muscular endurance, speed, and strength have been examined mainly using male subjects. The better physical performance of males in certain types of activities is well known, especially in those activities supported by anaerobic mechanisms (e.g., strength and muscle endurance). Several theories have been developed to explain these gender differences. Three characteristics of women contribute to their poorer performance on the WAnT: (1) their relatively inefficient skeletal configuration for certain physical demands; (2) their higher percentage of adipose tissue and lower fat-free mass; and (3) their lower peak lactic acid level in the blood and muscle following an all-out physical effort.

Data on a relatively small group of females (N = 70) ranging in age from 9 to 25 years were collected at the Wingate Institute (Inbar 1985). Figures 5.5 and 5.6 present the age-related values on these females obtained while performing the WAnT with legs and arms, respectively. For comparison, data on males in the same age groups were added.

As with males, there was a clear, positive correlation between age and performance on the WAnT. This relationship was stronger for the legs than for the arms (see figs. 5.5 and 5.6). In absolute terms, it was not possible to deter-

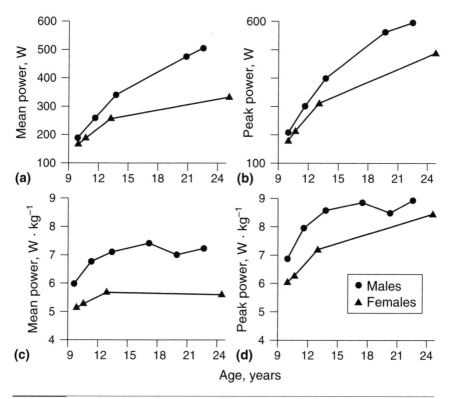

Figure 5.5 Leg anaerobic performance capacity and age in females (▲). Values for males (●) are given for comparison. Absolute values (W) for (a) mean power and (b) peak power. Relative values (W · kg⁻¹) are shown for (c) mean power and (d) peak power.

Adapted from Inbar 1985.

mine the age at which peak anaerobic performance occurred, since there was still a definite rise in absolute MP up to age 25 in the legs and up to 14 in the arms (see figs. 5.5a and b and 5.6a and b). When expressed in relative terms, however, the age at which peak anaerobic performance occurred in females was between 10 and 13 years, after which there was no further rise in MP · kg⁻¹ (see figs. 5.5c and d and 5.6c and d).

The gender difference in absolute MP in the legs was about 10% at the youngest age (9 years) and increased with age, reaching 20% at the age of 14 and 30% at the age of 25. When a gender comparison is made using a relative index (W · kg⁻¹), the gender gap remains the same or increases for both legs and arms.

Contrasting somewhat with the above data on Israeli youth are the results from a study by Blimkie et al. (1988) on Canadian school children, ages 14 to 16. When correction was made for lean arm volume, both MP and PP during arm cranking were marginally higher in the boys than in the girls. These gender-related differences became significant at 17 to 19 years of age. Maud

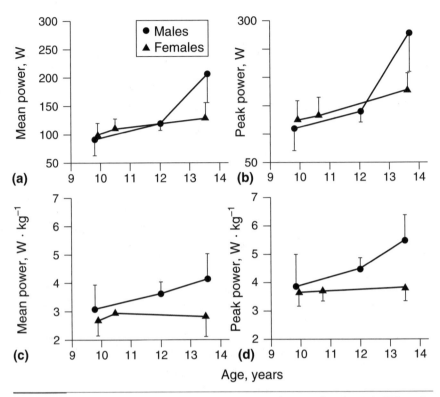

Figure 5.6   Arm anaerobic performance capacity and age in females (▲). Values for males (●) are given for comparison. Absolute values (W) are shown for (a) mean power and (b) peak power. Relative values (W · kg⁻¹) are shown for (c) mean power and (d) peak power.
Adapted from Inbar 1985.

and Shultz (1989) studied physically active American adults, ages 18 to 28. In absolute terms, the MP of the legs was 48% higher in males than in females. Calculated per kg body mass, the difference was 15%, but only 2% when calculated per kg fat-free mass. The respective differences for PP were 54%, 21%, and 7%. Thus, these two studies suggest strongly that the gender differences are mostly caused by muscle mass.

An interesting gender-related difference emerged from the results of a study on 21 women and 17 men, ages 30 to 40, who performed the arm and leg WAnT (Ben-Ari, Inbar, and Bar-Or 1978). As seen in figure 5.7, the men had higher scores when values were expressed per kg body mass. Because the men were heavier, these differences would have been even greater when expressed in absolute terms. The difference for the arms was much greater than for the legs: the MP with the arms was 52% higher in the men, while it was only 23% different with the legs. The respective differences for PP were 50% and 25%. Thus, it seems that muscle endurance of the arms is particularly low in women 30 to 40 years of age.

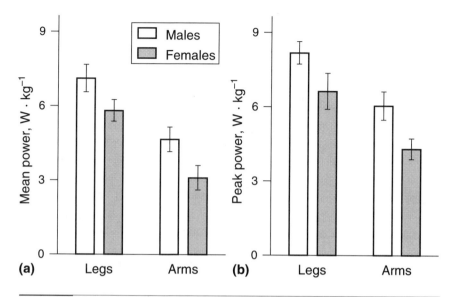

Figure 5.7 Leg and arm (a) mean power and (b) peak power relative to body mass ($W \cdot kg^{-1}$) generated during WAnT for males and females ages 30-40. All differences between males and females were significant ($p < 0.01$).
Reprinted from Ben-Ari, Inbar, and Bar-Or 1978.

Typical values of the various WAnT indices are presented in the appendix for healthy, untrained Israeli males and females for both leg and arm tests by age.

Sport Specialty

If the WAnT is a valid anaerobic test, then athletes who specialize in sprinting, jumping, or power events should score higher on the WAnT than endurance athletes. While several authors have reported PP and MP as part of an overall physiologic profile in various sport disciplines (e.g., Inbar and Bar-Or 1977; Inbar et al. 1989; Rhodes, Cox, and Quinney 1986; Smith, Quinney, and Steadward 1982), it is difficult to compare their findings because of different testing protocols, incompatible athletic proficiency, and unequal ages. Only a few studies are available in which various types of athletes were compared. Tharp et al. (1984) gave the WAnT to 21 girls and 18 boys (ages 10 to 17) who were members of an elite Nebraska track club. In the males, PP ($W \cdot kg^{-1}$) of the sprinters was 10.90 and significantly higher than that of the long-distance runners (9.94). A similar trend was found for MP (9.04 vs. 8.45 $W \cdot kg^{-1}$). The interspecialty differences were not significant among the females.

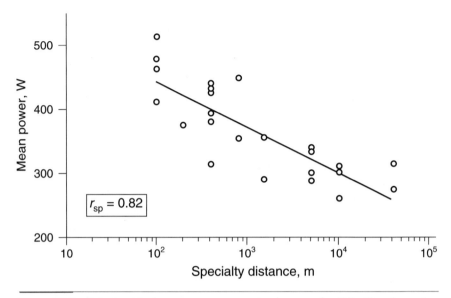

Figure 5.8   Relationship between mean power values on the WAnT and running specialty of male members of the Burmese national track team.
Reprinted from Bar-Or 1987.

Figure 5.8 plots the MP of male runners from the Burmese national track team versus the logarithm of their running specialty (Bar-Or 1987). It is apparent that the longer the running distance, the lower the WAnT score. The 10-km runners and the marathoners scored even lower than a comparable group of sedentary Burmese males.

Taunton, Maron, and Wilkinson (1981) compared the performance on the WAnT of young adult middle-distance and long-distance runners. While the PP ($W \cdot kg^{-1}$) of the middle-distance runners was significantly higher than that of the long-distance group, the difference in MP was not significant. It was also found that the power produced during the Margaria step-running test did not differ between the two running groups.

Studying the question of training specificity, Kohrt (1986) tested the aerobic and anaerobic performance of 11 male triathletes. Using 75 gm · kg⁻¹ for the legs and 50 gm · kg⁻¹ for the arms, the triathletes performed the WAnT on a Monark ergometer. As can be seen in table 5.2, their values on the leg and arm tests were within the range of values reported for active subjects, even though the values from the arm test were lower than would be expected from athletes who train with their arms.

In another study by Skinner and O'Connor (1987), 44 male athletes from several specialties performed the WAnT. See table 3.3 for a summary of those results. A pattern was apparent in which those athletes whose specialty was "anaerobic" in nature (power lifters) had significantly higher PP than the "aero-

Table 5.2   Mean Values ± (Standard Deviation) for Mean Power, Peak Power, and Fatigue Index Obtained From 11 Male Triathletes on the WAnT for Legs and Arms

Variable	Legs	Arms
Mean power (W · kg^{-1})	8.8 (0.8)	5.1 (0.5)
Peak power (W · kg^{-1})	11.2 (1.1)	6.7 (1.0)
Fatigue index (%)	38.1 (6.5)	41.7 (10.9)

bic" athletes (10-km runners and ultramarathoners), even when values were corrected for body mass. The scores of the gymnasts and wrestlers fell between the two extremes. Interestingly, there were no significant differences in MP among the groups (8.8 to 9.3 W · kg^{-1}). The 10-km runners and ultramarathoners had significantly lower rates of fatigue (26-33%) than the other three groups (43-47%). In other words, the "anaerobic athletes" had higher initial values but fatigued more rapidly, while the "aerobic" athletes had lower initial values but fatigued less rapidly. For reference (and comparison), WAnT values for legs and arms of some 371 male and female Israeli elite athletes specializing in various sport events are presented in figures 5.9 through 5.12. Test results are discussed below.

WAnT Values for Legs

Male Athletes.   Inbar (1985) found that the highest absolute values in W for leg MP and PP (indices more relevant for events not requiring the acceleration of body mass, e.g., rowing, cycling, swimming, and weight lifting) were found in rowers; the lowest values were found in long-distance runners. Sprinters and jumpers (power events) and players of water polo and ball games followed the rowers in absolute leg MP and PP, while cyclists, middle-distance runners, and martial-arts athletes were closer to the long-distance runners. As depicted in figure 5.9, the relative indices (MP in W · kg^{-1} and PP in W · kg^{-1}) are more relevant to sport events in which body mass is being moved, e.g., running, jumping, and ball games. These relative values show a similar pattern in that the rowers and the long-distance runners (not shown in fig. 5.9) showed the highest and the lowest values for PP, respectively. Rowers also had the highest values for MP. Similar rank order of the absolute and relative indices across sport disciplines implies a constant relationship between anaerobic performance capacity and body mass in those sport events studied.

In most athletes studied, leg MP was 20% to 30% lower than leg PP. Outstanding in this respect were athletes in martial arts who had leg MP values some 40% lower than their PP values, as well as rowers and cyclists who had rather similar MP and PP values. Such unique relationships between MP and PP may be significant to those events and a prerequisite for high achievement.

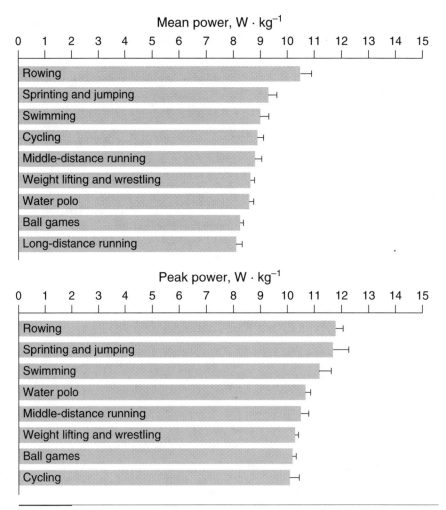

Figure 5.9   Leg anaerobic performance capacity (relative to body mass) of elite male Israeli athletes in various sport events.
Adapted from Inbar 1985.

In a study similar to the one carried out at the Wingate Institute, Di Prampero and colleagues in 1968 tested some 116 internationally ranked athletes during the Olympic Games in Mexico City using the Margaria step-running test (Di Prampero, Pinera-Limas, and Sassi 1970). These athletes competed in eight different sporting events and represented various countries. Although the events in the two studies were not identical and the athletic ability of the subjects in the latter study was obviously superior, the anaerobic performance ranking of each sport event shared by both was similar. An interstudy comparison revealed significantly higher absolute and relative anaerobic performance capacities among the subjects in the study by Di Prampero, Pinera-Limas, and Sassi (1970). A similar trend (i.e., higher absolute and relative values in the

Margaria test than in the WAnT) was reported by Ayalon, Inbar, and Bar-Or (1974) and by Cumming (1973). The higher mechanical power generated during the Margaria step-running test is attributable to the test's shorter and more explosive nature (about 0.5 sec) and the need to vertically lift the body mass.

In the study by Inbar (1985), the ranking among the different sports for the FI of the legs was similar to that of the relative peak power ($PP \cdot kg^{-1}$). As with untrained subjects, this was probably because of the close correlation between FI and the PP attained during the first few seconds of the WAnT. In several studies (Bar-Or et al. 1980; Campbell et al. 1979; Denis et al. 1990; Inbar, Kaiser, and Tesch 1981) significant positive correlation coefficients were found between FI and the percentage of fast-twitch (FT) muscle fibers. This relationship suggests that the percentage of FT fibers is relatively high among rowers, water polo players, sprinters, long jumpers, and high jumpers, and it is low among cyclists, middle-distance runners, and long-distance runners.

Female Athletes.   Intersport comparisons by Inbar (1985) of the absolute values for leg MP and PP in female athletes reveal the highest values for players of ball games, followed by athletes in power events, swimming, long-distance running, and gymnastics. A somewhat different pattern was evident in the relative indices ($MP \cdot kg^{-1}$ and $PP \cdot kg^{-1}$), with power events (mainly sprinting and jumping) producing the highest values and swimming the lowest. Gymnasts fell in between these two extremes (fig. 5.10).

Leg MP and PP values in female athletes tend to be 50% lower than those in male athletes. The corresponding gender difference in aerobic power ($\dot{V}O_2max$) is appreciably smaller (20-25%). This discrepancy is probably attributable to the greater relative contribution of muscle mass differences to anaerobic than to aerobic activities.

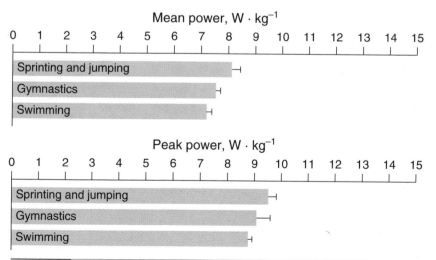

Figure 5.10   Leg anaerobic performance capacity (relative to body mass) of elite female Israeli athletes in various sport events.
Adapted from Inbar 1985.

Weight-adjusted MP and PP differences between male and female athletes are much smaller than the differences in absolute indices, especially in long-distance runners (0.5% and 4% in MP · kg^{-1} and PP · kg^{-1}, respectively), as well as in most other sports (10-25%). Adjusting the WAnT indices to lean body mass rather than to total body mass should bring the WAnT values of the two genders even closer.

WAnT Values for Arms

Male Athletes. For the WAnT arm indices, only athletes competing in those sporting events in which the arms play a dominant role were tested (fig. 5.11). As expected, rowers achieved the highest absolute and relative values. Among the three other events studied (ball games, swimming, and martial arts), arm anaerobic performance capacity was similar and significantly lower than that of the rowers.

In those male athletes in whom both leg and arm muscles were tested, MP and PP of the arms were about 60% to 65% of the values with their legs. A similar arm/leg ratio has been reported for the respective aerobic components. Outstanding in this relationship are the rowers, whose arm/leg anaerobic ratio was 80% to 85%. Such high arm/leg anaerobic ratios might represent an important factor in successful rowing performance and should be considered when designing training programs for these athletes.

Female athletes. Forty-five elite Israeli female swimmers, gymnasts, and handball players were tested for arm MP and PP. In absolute and relative terms,

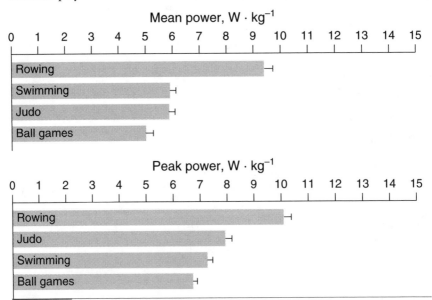

Figure 5.11 Arm anaerobic performance capacity (relative to body mass) of elite male Israeli athletes in various sport events.
Adapted from Inbar 1985.

the female handball players had the highest anaerobic performance, followed by the swimmers and the gymnasts. Figure 5.12 shows these values relative to body mass. Arm MP was some 25% lower than PP in the handball players and 20% lower in both the swimmers and the gymnasts. The arm/leg ratios in the sporting events studied were similar to those found in the male athletes (60-65%).

On the average, the arm anaerobic performance of female athletes was 25% to 30% lower than that of the male athletes. These gender-related differences in arm anaerobic performance are appreciably smaller than the respective differences in leg WAnT indices, which are about 50%.

Concluding Remarks

When interpreting these data, one should remember that in most events, top Israeli athletes are not on the same level as their internationally ranked counterparts. At least in some sport events studied, the sample size also was too small to establish reliable sport-specific reference values. In spite of this, these values generated from the largest and most diverse population of its kind can be a useful guide when anaerobic testing and interpretation in athletes of various sport disciplines are being considered.

Results of the studies cited in this chapter suggest that a "logical" relationship exists between performance in the WAnT and athletic specialty among males. Nevertheless, more studies are needed to firmly establish such relationships in the various sports, especially in female athletes.

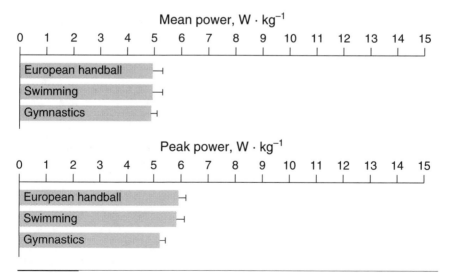

Figure 5.12   Arm anaerobic performance capacity (relative to body mass) of elite female Israeli athletes in various sport events.
Adapted from Inbar 1985.

Chronic Disease

Although originally used with able-bodied, healthy people, the WAnT can also be used to assess people with a chronic disease or a physical disability. The rationale for such an application has been that the factors limiting physical performance may be muscular or neurological rather than cardiorespiratory in certain groups of patients (Bar-Or 1986, 1993). Thus, testing their "peripheral" function may have diagnostic and prognostic value.

Methodological Considerations

Important questions remain about the feasibility and reliability of the WAnT when it is performed by people with a physical disability. We also still need to learn how to standardize the test for such people because of the marked variation in ability, fitness level, and active muscle mass. So far, only limited answers are available. These are based mainly on clinical and research experience from the Children's Exercise and Nutrition Centre in Hamilton, Canada, where children and adolescents with muscle dystrophy, muscle atrophy, cerebral palsy, spina bifida, cystic fibrosis, obesity, and anorexia nervosa have been routinely tested using the WAnT. When indicated, arm tests and/or leg tests were used.

In general, the WAnT is feasible for these youngsters, irrespective of their disability. In a study by Tirosh, Rosenbaum, and Bar-Or (1990), the arm WAnT and the leg WAnT were administered in duplicate to 66 girls and boys, ages 6 to 20, with spastic cerebral palsy (mostly tetraplegic), athetotic cerebral palsy, Duchenne muscular dystrophy, Becker muscular dystrophy, spinal muscular atrophy, congenital muscular atrophy, and myotonic dystrophy. Of these subjects, the arm test and the leg test were successfully performed by 92% and 61%, respectively. Those who could not perform either test were mostly children with Duchenne muscular dystrophy. In general, the leg test could be done by those children who could still walk with or without crutches, even when their disability was advanced. A similarly high feasibility has been reported for children and adolescents with cerebral palsy (Parker et al. 1992).

Many children with cerebral palsy (athetosis or spasticity) cannot keep their feet on the pedals throughout the test, even when stirrups are used. Excellent results were obtained when their shoes were taped to the pedal. Similar excellent results were obtained when the hands of some athetotic and spastic children were tied to the pedals during the arm test.

Patients with extreme muscle weakness sometimes find it impossible to complete a full pedal revolution with the legs or arms. Decreasing pedal length allows the rotation to occur at a smaller circumference. For example, with muscular dystrophy patients who may have some residual action of the smaller wrist muscles, but not of the elbow extensors, using a crank length of 12.5 cm is preferable to the usual length of 17.5 cm. Likewise, a child may be able to perform a full pedal revolution only if the seat is adjusted horizontally or vertically relative to the pedals.

Another practical consideration is the choice of an ergometer. Healthy children often reach a peak power output of 200 to 300 W with the legs and 150 to 250 W with the arms. Patients with moderate to advanced muscle weakness may have difficulty with 20 W. When these patients fatigue during the test, their power often drops to 5 W or less. Some children with advanced muscular dystrophy cannot pedal at zero resistance. Most ergometers available today cannot measure these low values because of poor resolution at the lower end of the scale and hard-to-measure power losses in the various components of the ergometer. Experience at McMaster University has shown that the Fleisch-Metabo ergometer provides only a partial solution. While it presumably can detect power outputs as low as 5 W, the actual muscle power generated is higher because of the forces needed to turn the pedals at zero resistance.

The optimal resistance for patients is not known because guidelines for resistance used with healthy persons are usually not applicable to patients with a disability in which the muscle mass/total body mass ratio is abnormal. This is particularly obvious for individuals with muscle atrophy or marked obesity, where using these tables would yield an overestimation of the optimal resistance because the muscle mass is smaller than expected from the total mass. So far there are no alternative, validated tables for such people. At the Children's Exercise and Nutrition Centre, patients try various resistances (in an ascending order) during the warm-up period and the highest one against which the child can "sprint" for 2-3 s is selected. While this practical solution still requires validation, designing a study to find optimal resistance for the disabled will be difficult because of the wide spectrum of diseases and levels of residual ability. Preliminary data (Van Mil et al. 1993) suggest that one can predict the optimal braking force for youth with cerebral palsy by first identifying a person's optimal force in the Force-Velocity Test (see chap. 2) and then administering a WAnT, using 65% of that force. Another possible approach would be to base the choice of an optimal force on the estimated lean arm (or leg) mass of the subject; this approach needs experimental validation.

The WAnT also has been used with elderly men and women, ages 54 to 84, with chronic obstructive lung disease (Bar-Or, Berman, and Salsberg 1992). The rationale for attempting the WAnT with such patients was that they had an irreversibly low respiratory function. Any rehabilitation/training program would therefore not improve their pulmonary function but might improve their "peripheral" function, i.e., leg muscle power and endurance. If so, then such an abbreviated WAnT could be used to assess the efficacy of rehabilitation programs.

The 18 older patients tested with an abbreviated (15-s) version of the WAnT all managed to complete the test twice without any subjective complaints or electrocardiographic and blood pressure aberrations during or after the test. While it is possible that these patients could have pedaled longer than 15 s, this was not attempted for ethical and safety reasons. Furthermore, considering the high correlation in younger subjects between performance on the full 30-s test and on shorter protocols (unpublished data from the Wingate laboratory; Vandewalle, Peres, and Monod 1987), it is likely that the 15-s version provided similar information.

Other groups at McMaster University (Jones and McCartney 1986; Markides et al. 1985) have used a 30-s isokinetic cycling test with healthy subjects up to 70 years of age, and they report no untoward effects. The same test was performed successfully by 33 middle-aged people with coronary artery disease (Jones and McCartney 1986), most of whom were tested within 3 months after sustaining a myocardial infarction and some following a coronary bypass operation.

A modified WAnT has also been performed by a group of 10 adult asthmatics (Inbar, Alvarez, and Lyons 1981). Their static and dynamic responses to this challenge were compared with those obtained during more conventional aerobic, steady-state exercise lasting 6-8 min; this is the classic protocol for exercise-induced asthma. As seen in figures 5.13 and 5.14, the submaximal aerobic exercise challenge produced no significant change in forced vital capacity (FVC), thoracic gas volume (T_{GV}), or specific conductance (S_{GA}). There was an expected significant rise in airway resistance (R_{AW}) and a drop in the forced-expiratory volume in 1 s (FEV_1) and the maximal mid-expiratory flow rate (MMEFR). Following the brief anaerobic exercise challenge of the WAnT, only the drop in MMEFR was significant. It was concluded that such

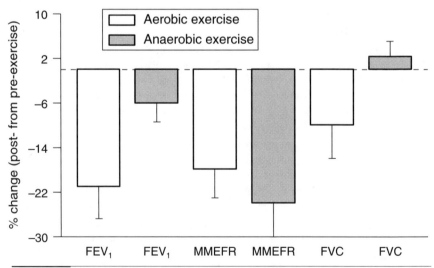

Figure 5.13   Changes in dynamic ventilatory function following submaximal aerobic exercise and short-maximal anaerobic (WAnT) exercise, as reflected by forced-expiratory volume in 1 s (FEV_1), maximal mid-expiratory flow rate (MMEFR), and forced vital capacity (FVC). Values are $\bar{X} \pm SEM$.
Reprinted from Inbar, Alvarez, and Lyons 1981.

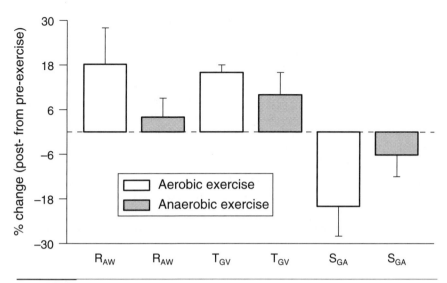

Figure 5.14   Changes in body plethysmographic measurements following submaximal aerobic exercise and short-maximal anaerobic (WAnT) exercise, as reflected by R_{AW} = airway resistance, T_{GV} = thoracic gas volume, and S_{GA} = specific conductance. Values are $\bar{X} \pm SEM$.
Reprinted from Inbar, Alvarez, and Lyons 1981.

a brief, all-out effort caused only peripheral bronchoconstriction, as indicated by the drop in MMEFR only, and may be used as a differential diagnostic test with asthmatic patients.

These observations suggest that such brief, high-intensity tasks are safe with elderly people, including those with respiratory or coronary disease. Nevertheless, special caution must be taken with such subjects, including a thorough warm-up and continuous blood pressure monitoring before and after the test and the electrocardiogram before, during, and after the test.

Reliability

In the study by Tirosh, Rosenbaum, and Bar-Or (1990), 58 disabled children completed the arm WAnT twice and 38 did the leg WAnT twice. The following test-retest correlation coefficients were found: arm PP = 0.94; arm MP = 0.98; and arm FI = 0.76. The respective coefficients for the legs were 0.96, 0.96, and 0.48. These high reliability coefficients for PP and MP are particularly impressive considering the subjects' severe disability. Similar means and standard deviations were found for each subgroup and for the total group, indicating good reproducibility. High test-retest correlation coefficients were also found in the elderly patients with chronic obstructive lung disease (0.89 for PP and 0.90 for MP).

Anaerobic Performance of Special Populations

As expected, muscle power and muscle endurance are low in children with neuromuscular disabilities. An example is shown in figure 5.15, in which the values of fourteen patients ages 8 to 14 for MP · kg^{-1} are compared with those of healthy children. Eleven patients had scores that were well below the values at two standard deviations below the mean value of the control subjects. While the maximal aerobic power of such patients is also low, the deficiency in local muscular performance seems even more pronounced. A similar pattern of low anaerobic power has been reported in children with cerebral palsy by Emons et al. (1992) and by Parker et al. (1992).

Based on clinical experience, the anaerobic exercise performance of obese children and adolescents does not seem to be lower than that seen in their nonobese counterparts. When calculated in absolute terms, PP and MP are

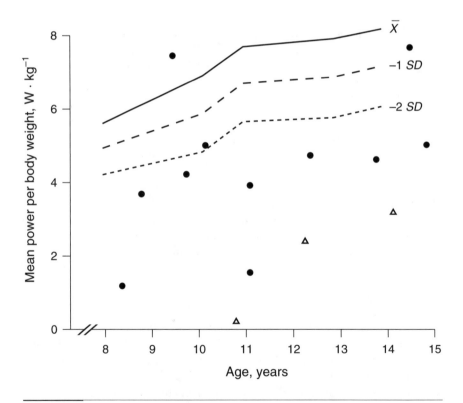

Figure 5.15 Leg anaerobic performance in pediatric neuromuscular disease. Eleven boys with cerebral palsy (●) and three with Duchenne muscular dystrophy (△) performed the Wingate anaerobic leg cycling test. Typical values for nonathletic healthy boys are from Bar-Or (1983).
Reprinted from Bar-Or 1986.

often higher than values seen in a matched nonobese population; this probably reflects the larger muscle mass of the obese. Obese children (particularly obese teenagers) are often taller and have a larger muscle mass than the nonobese. In contrast, when corrected for body mass, the muscle power generated by the obese is lower. Clinical and anecdotal observations from McMaster University suggest that there is no difference in the anaerobic performance of lean and obese children when the results are calculated per unit of muscle mass.

A reverse pattern is apparent in patients with cystic fibrosis, as they are often very lean. Their absolute PP and MP values are lower than those found in healthy cohorts. When calculated per kg body mass, however, these indices may reach average to above-average levels. As shown by Cabrera et al. (1993), anaerobic performance is related to the disease's severity. Children and adults with severe CF had considerably lower PP and MP (even when corrected for body mass) compared with patients whose disease was less advanced. No data are available on the muscle power per muscle mass in these patients.

Chapter 6

Conclusions and Challenges for Future Research

The protocol for the WAnT has undergone modifications and refinements since the prototype test was introduced in 1974. The use of a higher force to maximize power output represents a major change and is highly recommended. More research is needed, however, to pinpoint the optimal force for such subgroups as athletes of various specialties, children, the elderly, and the disabled. This should be done for leg and arm exercise alike.

Pedaling speed used to be monitored every 3-5 s. With the improvement in counting and recording techniques, it is now possible to increase the frequency of observations. This helps determine more precise (and higher) values for PP and the rate of fatigue.

Even when performed under nonstandardized, warm, or humid climatic conditions, as well as during mild to moderate hypohydration, ample evidence is now available that the WAnT is highly reliable, reproducible, and safe in various populations. While habituation, learning, and motivation may cause only small changes in PP and MP, warming up does improve performance. Research is needed to identify the most effective warm-up protocol. Preparation for the test and its execution should be carefully standardized.

Determining WAnT validity has been a challenge. Each of the observations reviewed in the preceding chapters sheds some light on whether the test indeed reflects anaerobic performance. While none of these experiments can be considered the definitive validation study, their cumulative message is that the WAnT is performed predominantly with anaerobic energy sources and that it taxes the anaerobic energy pathways. High scores on this test reflect high anaerobic performance. It would be presumptuous, however, to expect this test (or any other physiologic test) to predict performance in high-power sport events. In such events, skill, tactics, and fitness components other than anaerobic fitness may be just as important. Sport-specific anaerobic tests (e.g., a tread-mill test for runners) also may yield more information that is directly appli-cable to the movements involved in a particular sport.

A standardized, generally accepted system for classifying anaerobic ability is needed. Skinner and Morgan (1985) proposed such a modified system based on characteristics and limitations of the body's energy systems. There were two aerobic classes, two anaerobic classes, and one aerobic/anaerobic class in this system. More research is needed for classifying different types of activity.

As mentioned in chapter 3, Skinner and O'Connor (1987) did a cross-sectional study on athletes who performed the five types of activities in the proposed classification system (power lifters, gymnasts, wrestlers, 10-km runners, and ultramarathoners) in the continuum from high anaerobic to high aerobic ability. Although the 5-s PP separated the anaerobic power lifters from the aerobic runners, there were no differences in the 30-s MP of the different athletes. It appears that a 5-s test and a 30-s test measure different anaerobic abilities in athletes. Given that these tests were done at a resistance of 75 $g \cdot kg^{-1}$ on a Monark ergometer and that this is lower than optimal resistance for all types of athletes (especially the power athletes) or for both 5-s *and* 30-s anaerobic tests, any discussion of the optimal duration for an anaerobic test should also consider the optimal resistance and the anaerobic ability that one wishes to measure. More research is needed.

For theoretical and practical reasons, a 30-s test has been developed. Fur-ther research may show that a somewhat longer duration yields different in-formation, particularly on aerobic abilities and the fatigue pattern. Conversely, a shorter version may be indicated for use with disabled and elderly popula-tions or whenever PP is the variable in question.

Different types of anaerobic ability and many kinds of anaerobic tests obvi-ously exist. As stated by Vandewalle, Peres, and Monod (1987), there is no single anaerobic test that measures the different components of anaerobic metabolism equally well. Thus, more research on the biochemical and neural events associated with each test is warranted so that we can better understand anaerobic metabolism and performance. Specifically, research is needed to better understand what biochemical events are reflected by PP and MP. An-other area of investigation concerns the metabolic and neurologic correlates of certain fatigue curves. With better understanding of general anaerobic abil-ity, more sport-specific tests can be developed in the laboratory and then modi-fied and applied to the athlete.

Much more research is warranted in the field of muscle performance and disability. One potential area is the use of the WAnT as a tool for assessing the functional corollary of one's nutritional status. In recent years, relationships have been described between the function of the abductor pollicis muscle (activated by electrical stimulation of the ulnar nerve) and the nutritional sta-tus of people with gastrointestinal disorders (Lopes et al. 1982) and anorexia nervosa (Russell et al. 1983). Muscle force, maximal relaxation rate, and muscle endurance (force loss over 30 s) were taken as the functional criteria. Few data are available, however, regarding nutritional status and the function of larger muscle groups, as assessed by the WAnT (Hanning et al. 1993).

Another challenge is to establish norms for the relationship *within a given individual* between peak anaerobic and peak aerobic power (e.g., Blimkie, Roche, and Bar-Or 1986; Falk and Bar-Or 1993). These norms may depend on age and gender, and they vary among athletic specialties. Deviation from such norms may suggest to the coach (or the therapist) how to modify a training (or rehabilitation) regimen. Likewise, one can determine the ratio between a person's arm performance and leg performance on the WAnT. Using normative data for such a ratio, one could then suggest changes in the training regimen. Such information would be relevant to such athletic specialties as rowing, wrestling, skiing, or swimming.

Appendix

Typical WAnT Values for Healthy, Untrained Israeli Males and Females, Ages 8 to 45

The values offered here are based on data obtained at the Wingate Institute over several years. These "typical" values should not be considered norms, as that would have required using large representative samples of various populations. Such data based on age, gender, health status, etc., are not available.

Typical Values (Legs) for Healthy, Untrained Israeli Males

Table 1. Absolute Mean Power (MP), W

Table 2. Absolute Peak Power (PP), W

Table 3. Relative Mean Power (MP $\cdot$ kg^{-1}), W $\cdot$ kg^{-1}

Table 4. Relative Peak Power (PP $\cdot$ kg^{-1}), W $\cdot$ kg^{-1}

Typical Values (Arms) for Healthy, Untrained Israeli Males

Table 5. Absolute Mean Power (MP), W

Table 6. Absolute Peak Power (PP), W

Table 7. Relative Mean Power (MP $\cdot$ kg^{-1}), W $\cdot$ kg^{-1}

Table 8. Relative Peak Power (PP $\cdot$ kg^{-1}), W $\cdot$ kg^{-1}

Typical Values (Legs) for Healthy, Untrained Israeli Females

Table 9. Absolute Mean Power (MP), W

Table 10. Absolute Peak Power (PP), W

Table 11. Relative Mean Power (MP $\cdot$ kg^{-1}), W $\cdot$ kg^{-1}

Table 12. Relative Peak Power (PP $\cdot$ kg^{-1}), W $\cdot$ kg^{-1}

Typical Values (Arms) for Healthy, Untrained Israeli Females

Table 1. Typical Values (Legs) for Healthy, Untrained Israeli Males
Absolute Mean Power (MP), W

Age: Category	<10	10-12	12-14	14-16	16-18	18-25	25-35	>35	% of cases in normal distribution
Very poor	115-145	100-169	203-263	140-234	278-342	302-380	371-449	359-412	11
Poor	146-161	169-204	263-292	234-280	342-375	380-418	449-488	412-449	12
Below average	161-176	204-239	292-321	280-327	375-407	418-456	488-526	449-480	17
Average	176-192	239-274	321-350	327-373	407-439	456-494	527-565	480-510	20
Good	192-207	274-308	350-380	373-420	439-472	495-533	565-604	510-540	17
Very good	207-223	309-343	380-409	420-467	472-504	533-571	604-642	540-570	12
Excellent	223-254	343-413	409-467	467-599	504-569	571-649	642-719	570-630	11
$\bar{X}$	184.12	256.24	335.79	350.05	423.21	475.38	545.66	494.70	
SD	30.87	69.69	58.52	92.90	64.70	76.35	77.20	60.36	
N	8	89	16	5	17	21	16	5	

Table 2. Typical Values (Legs) for Healthy, Untrained Israeli Males
Absolute Peak Power (PP), W

Age: Category	<10	10-12	12-14	14-16	16-18	18-25	25-35	>35	% of cases in normal distribution
Very poor	134-170	140-196	242-318	186-288	333-412	339-440	450-566	389-513	11
Poor	171-187	196-238	318-353	288-339	412-451	441-492	566-623	514-576	12
Below average	187-204	238-280	353-389	339-392	451-489	493-544	623-680	576-638	17
Average	204-221	280-323	389-425	392-444	490-528	544-596	680-737	638-700	20
Good	221-238	323-365	425-461	444-497	528-567	596-648	738-795	700-762	17
Very good	238-254	365-407	461-496	497-549	567-606	648-700	795-852	762-825	12
Excellent	255-288	408-490	496-568	549-651	606-685	700-802	852-969	825-950	11
$\bar{X}$	212.50	301.56	407.00	418.25	508.96	570.41	708.92	669.01	
SD	33.40	84.75	71.43	104.45	77.79	103.94	114.37	124.47	
N	8	89	16	5	17	21	16	5	

Table 3. Typical Values (Legs) for Healthy, Untrained Israeli Males
Relative Mean Power (MP · kg⁻¹), W · kg⁻¹

Age: Category	<10	10-12	12-14	14-16	16-18	18-25	25-35	>35	% of cases in normal distribution
Very poor	3.7-4.7	4.7-5.6	4.9-5.9	4.7-5.5	5.9-6.5	5.1-6.0	6.1-6.7	4.7-5.5	11
Poor	4.7-5.2	5.6-6.1	5.9-6.4	5.5-5.9	6.5-6.9	6.0-6.4	6.7-6.9	5.5-5.9	12
Below average	5.2-5.7	6.1-6.6	6.4-6.8	5.9-6.3	6.9-7.2	6.4-6.9	7.0-7.2	5.9-6.3	17
Average	5.8-6.3	6.6-7.0	6.8-7.3	6.3-6.7	7.2-7.5	6.9-7.3	7.3-7.5	6.3-6.7	20
Good	6.3-6.8	7.0-7.5	7.3-7.8	6.7-7.1	7.5-7.9	7.3-7.7	7.6-7.8	6.7-7.1	17
Very good	6.8-7.3	7.5-8.0	7.8-8.2	7.1-7.5	7.9-8.2	7.7-8.2	7.8-8.2	7.1-7.5	12
Excellent	7.3-8.4	8.0-8.9	8.2-9.2	7.5-8.5	8.2-9.0	8.2-9.0	8.2-8.7	7.5-8.3	11
$\bar{X}$	6.01	6.80	7.08	6.46	7.36	7.07	7.40	6.52	
SD	1.04	0.95	0.93	0.80	0.66	0.86	0.59	0.81	
N	8	89	16	5	17	21	16	5	

Table 4. Typical Values (Legs) for Healthy, Untrained Israeli Males

Relative Peak Power (PP · kg⁻¹), W · kg⁻¹

Age: Category	<10	10-12	12-14	14-16	16-18	18-25	25-35	>35	% of cases in normal distribution
Very poor	4.2-5.5	5.4-6.6	6.2-7.3	5.7-6.6	7.0-7.8	5.4-6.8	7.8-8.6	5.6-7.0	11
Poor	5.5-6.0	6.6-7.1	7.3-7.8	6.6-7.1	7.8-8.3	6.8-7.5	8.6-9.0	7.0-7.7	12
Below average	6.1-6.6	7.1-7.7	7.8-8.3	7.1-7.5	8.3-8.6	7.5-8.1	9.0-9.4	7.7-8.4	17
Average	6.7-7.2	7.7-8.3	8.3-8.8	7.6-8.0	8.7-9.1	8.2-8.8	9.4-9.8	8.4-9.1	20
Good	7.3-7.8	8.3-8.8	8.8-9.3	8.0-8.5	9.1-9.5	8.8-9.5	9.8-10.2	9.2-9.9	17
Very good	7.9-8.4	8.8-9.4	9.3-9.8	8.5-8.9	9.5-9.9	9.5-10.2	10.2-10.6	9.9-10.6	12
Excellent	8.4-9.6	9.4-10.6	9.9-10.9	8.9-9.8	9.9-10.8	10.2-11.6	10.6-11.3	10.6-12.0	11
$\bar{X}$	6.95	7.98	8.56	7.78	8.86	8.50	9.59	8.79	
SD	1.19	1.15	1.03	0.91	0.84	1.38	0.78	1.44	
N	8	89	16	5	17	21	16	5	

Table 5. Typical Values (Arms) for Healthy, Untrained Israeli Males
Absolute Mean Power (MP), W

Age: Category	<10	10-12	12-14	14-16	16-18	18-25	25-35	>35	% of cases in normal distribution
Very poor	22-47	80-98	82-137	120-179	137-197	145-218	234-274	190-249	11
Poor	47-65	98-107	137-165	179-207	197-227	218-255	274-294	249-279	12
Below average	65-83	107-116	165-192	207-236	227-258	255-292	294-315	279-308	17
Average	83-100	116-124	192-220	236-264	258-288	292-329	315-335	308-337	20
Good	100-118	124-133	220-248	264-293	288-318	329-366	335-355	337-366	17
Very good	118-136	133-142	248-276	293-321	318-349	366-403	355-375	366-395	12
Excellent	136-161	142-159	276-333	321-380	349-409	403-477	375-415	395-454	11
$\bar{X}$	91.43	120.11	206.26	250.02	272.86	310.47	324.60	322.15	
SD	35.26	17.36	55.36	36.90	60.70	73.82	40.30	58.18	
N	3	22	28	6	17	19	18	9	

Table 6. Typical Values (Arms) for Healthy, Untrained Israeli Males
Absolute Peak Power (PP), W

Age: Category	<10	10-12	12-14	14-16	16-18	18-25	25-35	>35	% of cases in normal distribution
Very poor	28-60	93-115	71-162	190-252	165-256	212-311	277-341	238-317	11
Poor	60-80	115-126	162-207	252-284	256-301	311-360	341-373	317-356	12
Below average	80-101	126-137	207-253	284-316	301-347	360-409	373-405	356-394	17
Average	101-122	137-148	253-298	316-348	347-393	409-458	405-437	394-433	20
Good	122-143	148-159	298-343	348-379	393-438	458-507	437-469	433-471	17
Very good	143-164	159-171	343-389	379-411	438-484	507-556	469-501	471-510	12
Excellent	164-205	171-192	389-473	411-473	484-575	556-658	501-565	510-589	11
$\bar{X}$	111.67	142.61	275.26	331.78	369.94	433.28	420.94	413.43	
SD	41.66	22.40	90.75	63.31	91.33	98.08	64.04	77.30	
N	3	22	28	6	17	19	18	9	

Table 7. Typical Values (Arms) for Healthy, Untrained Israeli Males
Relative Mean Power (MP · kg⁻¹), W · kg⁻¹

Age: Category	<10	10-12	12-14	14-16	16-18	18-25	25-35	>35	% of cases in normal distribution
Very poor	1.5-2.2	2.7-3.1	2.4-3.2	2.9-3.7	3.3-3.9	2.8-3.6	3.7-4.0	2.5-3.3	11
Poor	2.2-2.6	3.2-3.4	3.2-3.6	3.7-4.0	3.9-4.2	3.6-4.0	4.0-4.2	3.3-3.7	12
Below average	2.6-3.0	3.4-3.6	3.6-4.0	4.1-4.4	4.2-4.5	4.0-4.5	4.2-4.4	3.7-4.1	17
Average	3.0-3.4	3.6-3.8	4.0-4.4	4.5-4.8	4.6-4.9	4.5-4.9	4.4-4.5	4.1-4.5	20
Good	3.4-3.8	3.8-4.1	4.4-4.8	4.8-5.2	4.9-5.2	4.9-5.4	4.5-4.7	4.5-4.9	17
Very good	3.8-4.2	4.1-4.3	4.8-5.2	5.2-5.5	5.2-5.5	5.4-5.8	4.7-4.9	4.9-5.3	12
Excellent	4.2-4.7	4.3-4.9	5.2-6.0	5.5-6.3	5.5-6.2	5.8-6.6	4.9-5.2	5.3-6.1	11
$\bar{X}$	3.20	3.73	4.16	4.60	4.72	4.69	4.46	4.29	
SD	0.76	0.46	0.80	0.74	0.65	0.86	0.34	0.78	
N	3	22	28	6	17	18	18	9	

Table 8. Typical Values (Arms) for Healthy, Untrained Israeli Males
Relative Peak Power (PP · kg⁻¹), W · kg⁻¹

Age:\nCategory	<10	10-12	12-14	14-16	16-18	18-25	25-35	>35	% of cases in normal distribution
Very poor	1.9-2.8	3.5-3.9	2.6-3.9	4.5-5.2	4.0-5.0	4.4-5.4	4.6-5.1	3.3-4.3	11
Poor	2.8-3.2	3.9-4.1	3.9-4.5	5.2-5.5	5.1-5.6	5.4-5.8	5.1-5.4	4.3-4.8	12
Below average	3.2-3.7	4.1-4.3	4.5-5.2	5.5-5.9	5.6-6.1	5.8-6.3	5.4-5.6	4.8-5.3	17
Average	3.7-4.1	4.3-4.5	5.2-5.8	5.9-6.2	6.1-6.7	6.3-6.8	5.6-5.9	5.3-5.7	20
Good	4.1-4.6	4.5-4.8	5.8-6.4	6.2-6.6	6.7-7.2	6.8-7.3	5.9-6.2	5.7-6.2	17
Very good	4.6-5.0	4.8-4.9	6.4-7.1	6.6-6.9	7.2-7.7	7.3-7.7	6.2-6.4	6.2-6.7	12
Excellent	5.0-6.0	4.9-5.4	7.1-8.3	6.9-7.6	7.7-8.8	7.7-8.6	6.4-7.9	6.7-7.7	11
$\bar{X}$	3.92	4.40	5.50	6.05	6.39	6.53	5.76	5.50	
SD	0.90	0.42	1.25	0.68	1.07	0.93	0.50	0.98	
N	3	22	28	6	17	19	18	9	

Table 9. Typical Values (Legs) for Healthy, Untrained Israeli Females
Absolute Mean Power (MP), W

Age: Category	<10	10-12	12-14	14-16	16-18	18-25	25-35	>35	% of cases in normal distribution
Very poor	151-158	112-139	119-182			177-247			11
Poor	158-162	139-152	182-213			247-282			12
Below average	162-165	152-178	213-244			282-317			17
Average	165-169	178-191	244-275			317-352			20
Good	169-172	191-205	275-306			352-387			17
Very good	172-176	205-218	306-338			387-422			12
Excellent	176-183	218-244	338-400			422-492			11
$\bar{X}$	167.1	184.7	259.7			334.3			
SD	47.4	20.1	62.2			58.8			
N	3	9	13			18			

Table 10. Typical Values (Legs) for Healthy, Untrained Israeli Females
Absolute Peak Power (PP), W

Age: Category	<10	10-12	12-14	14-16	16-18	18-25	25-35	>35	% of cases in normal distribution
Very poor	145-165	106-158	138-224			319-400			11
Poor	165-175	158-184	224-267			400-441			12
Below average	175-185	184-210	267-310			441-482			17
Average	185-195	210-235	310-353			482-524			20
Good	195-205	235-261	353-396			524-565			17
Very good	205-215	261-287	396-439			565-605			12
Excellent	215-235	287-339	439-525			606-686			11
$\bar{X}$	190.3	220.5	331.3			503.0			
SD	21.3	51.7	85.9			93.2			
N	3	9	13			18			

Table 11. Typical Values (Legs) for Healthy, Untrained Israeli Females
Relative Mean Power (MP · kg⁻¹), W · kg⁻¹

Age: Category	<10	10-12	12-14	14-16	16-18	18-25	25-35	>35	% of cases in normal distribution
Very poor	4.65-4.84	2.98-3.99	4.00-4.76			4.31-4.90			11
Poor	4.85-4.94	4.00-4.51	4.77-5.14			4.91-5.20			12
Below average	4.95-5.04	4.52-5.03	5.15-5.52			5.21-5.50			17
Average	5.05-5.15	5.04-5.56	5.53-5.90			5.51-5.80			20
Good	5.15-5.24	5.57-6.07	5.91-6.28			5.81-6.10			17
Very good	5.25-5.34	6.08-6.59	6.29-6.65			6.11-6.40			12
Excellent	5.35-5.54	6.60-7.64	6.66-7.43			6.41-7.00			11
$\bar{X}$	5.15	5.31	5.72			5.66			
SD	0.37	1.08	0.76			0.59			
N	3	9	13			18			

Table 12. Typical Values (Legs) for Healthy, Untrained Israeli Females
Relative Peak Power (PP · kg⁻¹), W · kg⁻¹

Age: Category	<10	10-12	12-14	14-16	16-18	18-25	25-35	>35	% of cases in normal distribution
Very poor	3.85-4.84	2.98-4.42	4.80-5.88			6.28-7.27			11
Poor	4.85-5.34	4.43-5.17	5.89-6.43			7.28-7.77			12
Below average	5.35-5.84	5.18-5.92	6.44-6.97			7.78-8.27			17
Average	5.85-6.34	5.93-6.67	6.98-7.52			8.28-8.78			20
Good	6.35-6.84	6.68-7.42	7.53-8.06			8.79-9.28			11
Very good	6.85-7.34	7.43-8.17	8.07-8.61			9.29-9.78			12
Excellent	7.35-8.34	8.18-9.68	8.62-9.71			9.79-10.78			7
$\bar{X}$	6.11	6.34	7.26			8.53			
SD	1.02	1.53	1.09			1.07			
N	3	9	13			18			

Table 13. **Typical Values (Arms) for Healthy, Untrained Israeli Females**
Absolute Mean Power (MP), W

Age: Category	<10	10-12	12-14	14-16	16-18	18-25	25-35	>35	% of cases in normal distribution
Very poor	50-73	62-83	64-93						11
Poor	73-84	84-94	93-108						12
Below average	84-96	94-105	108-122						17
Average	96-107	105-116	122-137						20
Good	107-118	116-126	137-151						17
Very good	118-130	126-137	151-165						12
Excellent	130-153	137-158	165-194						11
$\bar{X}$	101.25	110.10	129.4						
SD	22.92	21.35	28.89						
N	5	8	15						

Table 14. Typical Values (Arms) for Healthy, Untrained Israeli Females
Absolute Peak Power (PP), W

Age: Category	<10	10-12	12-14	14-16	16-18	18-25	25-35	>35	% of cases in normal distribution
Very poor	53-86	55-89	110-140						11
Poor	86-102	89-106	140-155						12
Below average	103-119	106-124	155-170						17
Average	119-135	124-141	170-184						20
Good	135-152	141-159	184-199						17
Very good	152-168	159-176	199-214						12
Excellent	168-201	176-211	214-244						11
$\bar{X}$	127.2	132.52	177.04						
SD	32.86	35.02	29.65						
N	5	8	15						

Table 15. Typical Values (Arms) for Healthy, Untrained Israeli Females
Relative Mean Power (MP · kg⁻¹), W · kg⁻¹

Age: Category	<10	10-12	12-14	14-16	16-18	18-25	25-35	>35	% of cases in normal distribution
Very poor	1.4-2.0	2.4-2.6	1.2-2.0						11
Poor	2.0-2.4	2.6-2.7	2.0-2.3						12
Below average	2.4-2.7	2.7-2.8	2.3-2.7						17
Average	2.7-3.0	2.8-3.0	2.7-3.0						20
Good	3.0-3.3	3.0-3.1	3.0-3.4						17
Very good	3.3-3.6	3.1-3.2	3.4-3.8						12
Excellent	3.6-4.3	3.2-3.5	3.8-4.5						11
$\bar{X}$	2.84	2.91	2.86						
SD	0.64	0.25	0.72						
N	5	8	15						

Table 16. Typical Values (Arms) for Healthy, Untrained Israeli Females
Relative Peak Power (PP · kg⁻¹), W · kg⁻¹

Age: Category	<10	10-12	12-14	14-16	16-18	18-25	25-35	>35	% of cases in normal distribution
Very poor	2.6-3.1	3.0-3.3	2.6-3.2						11
Poor	3.1-3.4	3.3-3.5	3.2-3.4						12
Below average	3.4-3.6	3.5-3.7	3.5-3.7						17
Average	3.6-3.9	3.7-3.9	3.7-4.0						20
Good	3.9-4.1	4.0-4.2	4.0-4.3						17
Very good	4.1-4.4	4.2-4.4	4.3-4.6						12
Excellent	4.4-4.9	4.4-4.8	4.6-5.1						11
$\bar{X}$	3.73	3.81	3.86						
SD	0.52	0.42	0.55						
N	5	8	15						

References

Åstrand, P.O. 1952. *Experimental studies of physical working capacity in relation to sex and age.* Copenhagen: Munksgaard.

Ayalon, A., O. Inbar, and O. Bar-Or. 1974. Relationships among measurements of explosive strength and anaerobic power. In *International series on sports sciences.* 1: *Biomechanics IV,* ed. R.C. Nelson and C.A. Morehouse, 527-32. Baltimore: University Park Press.

Bar-Or, O. 1983. *Pediatric sports medicine for the practitioner: From physiologic principles to clinical applications,* 2-16. New York: Springer-Verlag.

————. 1986. Pathophysiological factors which limit the exercise capacity of the sick child. *Med. Sci. Sports Exerc.* 18: 276-82.

————. 1987. The Wingate Anaerobic Test—an update on methodology, reliability, and validity. *Sports Med.* 4: 381-94.

————. 1993. Noncardiopulmonary pediatric exercise tests. In *Pediatric laboratory exercise testing,* ed. T.W. Rowland, 165-85. Champaign, IL: Human Kinetics.

Bar-Or, O., L. Berman, and A. Salsberg. 1992. An abbreviated Wingate anaerobic test for women and men of advanced age. *Med. Sci. Sports Exerc.* 24: S22.

Bar-Or, O., R. Dotan, and O. Inbar. 1977. A 30-second all-out ergometric test: Its reliability and validity for anaerobic capacity. *Isr. J. Med. Sci.* 13: 326.

Bar-Or, O., R. Dotan, O. Inbar, A. Rotstein, J. Karlsson, and P. Tesch. 1980. Anaerobic capacity and muscle fiber type distribution in man. *Int. J. Sports Med.* 1: 89-92.

Bar-Or, O., and O. Inbar. 1978. Relationships among anaerobic capacity, sprint, and middle distance running of school children. In *Physical fitness assessment,* ed. R.J. Shephard and H. Lavallée, 142-47. Springfield, IL: Charles C Thomas.

Bar-Or, O., and L.D. Zwiren. 1975. Maximal oxygen consumption test during arm exercise: Reliability and validity. *J. Appl. Physiol.* 38: 424-26.

Baxter, C., and T. Reilly. 1983. Influence of time of day on all-out swimming. *Br. J. Sports Med.* 17: 122-27.

Bedu, M., N. Fellmann, H. Spielvogel, G. Falgairette, E. van Praagh, and J. Coudert. 1991. Force-velocity and 30-s Wingate tests in boys at high and low altitudes. *J. Appl. Physiol.* 70: 1031-37.

Ben-Ari, E., O. Inbar, and O. Bar-Or. 1978. The anaerobic capacity and maximal anaerobic power of 30-40 year old men and women. *Proceedings of the fifth international symposium of kinanthropometry and ergometry,* 427-33. Quebec: Pelican.

Bergh, U. 1980. Human power at subnormal body temperature. *Acta Physiol. Scand. (Suppl.)* 478.

Bergström, J., R. Harris, E. Hultman, and L. Nordesjö. 1971. Energy-rich phosphagens in dynamic and static work. In *Muscle metabolism during exercise*, ed. B. Pernow and B. Saltin, 341-56. New York: Plenum Press.

Blimkie, C.J.R., P. Roche, and O. Bar-Or. 1986. Concept of anaerobic to aerobic power ratio in pediatric health and disease. In *Children and exercise XII*, ed. J. Rutenfranz, 31-37. Champaign, IL: Human Kinetics.

Blimkie, C.J.R., P. Roche, J. Hay, and O. Bar-Or. 1988. Anaerobic power of arms in teenage boys and girls: Relationship to lean tissue. *Eur. J. Appl. Physiol.* 57: 677-83.

Borg, G. 1962. Physical performance and perceived exertion. Thesis. Lund: Gleerup.

Borg, G., E.G. Edstrom, and G. Marklund. 1971. *A bicycle ergometer for physiological and psychological studies*. Report no. 24, Institute for Applied Psychology, University of Stockholm.

Bouchard, C., M. Boulay, J.-A. Simoneau, G. Lortie, and L. Pérusse. 1988. Heredity and trainability of aerobic and anaerobic performances: An update. *Sports Med.* 5: 69-73.

Bouchard, C., A.W. Taylor, J.-A. Simoneau, and S. Dulac. 1991. Testing anaerobic power and capacity. In *Physiological testing of the high-performance athlete*, ed. J.D. MacDougall, H.A. Wegner, and H.J. Green, 175-221. Champaign, IL: Human Kinetics.

Burke, R.E., D.N. Levine, and F.E. Zajac. 1971. Mammalian motor units: Physiological-histochemical correlation in three types in cat gastrocnemius. *Science* 174: 709-12.

Cabrera, M.E., M.D. Lough, C.F. Doershuk, and G.A. DeRiva. 1993. Anaerobic performance—assessed by the Wingate test—in patients with cystic fibrosis. *Pediatr. Exerc. Sci.* 5: 78-87.

Cabri, J., B. DeWitte, J. Clarys, T. Reilly, and D. Strass. 1988. Circadian variation in blood pressure responses to muscular exercise. *Ergonomics* 31: 1559-65.

Campbell, C., A. Bonen, R. Kirby, and A. Belcastro. 1979. Muscle fiber composition and performance capacities of women. *Med. Sci. Sports Exerc.* 11: 260-65.

Chaloupecky, R. 1972. An evaluation of the validity of selected tests for predicting maximal oxygen uptake. Unpublished doctoral dissertation. Stillwater: Oklahoma State University.

Crielaard, J., and F. Pirnay. 1981. Anaerobic and aerobic power of top athletes. *Eur. J. Appl. Physiol.* 47: 295-300.

Cumming, G.R. 1973. Correlation of athletic performance and aerobic power in 12-17 year-old children with bone age, calf muscle, total body potassium, heart volume and two indices of anaerobic power. In *Pediatric work physiology*, ed. O. Bar-Or, 109-35. Netanya, Israel: Wingate Institute.

Cumming, G.R., L. Hastman, J. McCort, and S. McCullough. 1980. High serum lactates do occur in young children after maximal work. *Int. J. Sports Med.* 1: 66-69.

Cunningham, D., and J. Faulkner. 1969. The effect of training on aerobic and anaerobic metabolism during a short exhaustive run. *Med. Sci. Sports Exerc.* 1: 65-69.

Davies, C.T.M. 1971. Human power output in exercise of short duration in relation to body size and composition. *Ergonomics* 14: 245-56.

Davies, C.T.M., C. Barnes, and S. Godfrey. 1972. Body composition and maximal exercise performance in children. *Hum. Biol.* 44: 195-214.

Davies, C.T.M., and R. Rennie. 1968. Human power output. *Nature* 217: 770.

De Bruyn-Prévost, P. 1975. Essai de mise au point d'une epreuve anaérobie sur bicyclette ergométrique. *Med. du Sport* 49: 202-6.

De Bruyn-Prévost, P., and F. Lefèbvre. 1980. The effects of various warming up intensities and durations during a short maximal anaerobic exercise. *Eur. J. Appl. Physiol.* 43: 101-7.

Denis, C., M.T. Linossier, D. Dormois, R. Fouquet, A. Geyssant, J.R. LaCour, and O. Inbar. 1990. Specific responses of the Wingate Anaerobic Test to sprint versus endurance training: Effects of the adjustment of the load. In *Proceedings of the Maccabiah-Wingate International Congress, Life Sciences*, ed. G. Tenenbaum and D. Eiger, 9-17. Netanya, Israel: Wingate Institute.

Di Prampero, P.E., and P. Cerretelli. 1969. Maximal muscular power (aerobic and anaerobic) in African natives. *Ergonomics* 12: 51-59.

Di Prampero, P.E., F. Pinera-Limas, and G. Sassi. 1970. Maximal muscular power (aerobic and anaerobic) in 116 athletes performing at the XIXth Olympic Games in Mexico. *Ergonomics* 13: 665-74.

Dotan, R., and O. Bar-Or. 1980. Climatic heat stress and performance in the Wingate Anaerobic Test. *Eur. J. Appl. Physiol.* 44: 237-43.

———. 1983. Load optimization for the Wingate Anaerobic Test. *Eur. J. Appl. Physiol.* 51: 409-17.

Emons, H.J.G., D.C. Groenenboom, Y.I. Burggraaf, T.L.E. Janssen, and M.A. Van Baak. 1992. Wingate Anaerobic Test in children with cerebral palsy. In *Children and exercise XVI*, ed. J. Coudert and E. Van Praagh, 187-89. Paris: Mason.

Eriksson, B.O. 1980. Muscle metabolism in children: A review. *Acta Paediatr. Scand. (Suppl.)* 283: 20-27.

Eriksson, B.O., P.D. Gollnick, and B. Saltin. 1973. Muscle metabolism and enzyme activities after training in boys 11-13 years old. *Acta Physiol. Scand.* 87: 485-97.

Eriksson, B.O., J. Karlsson, and B. Saltin. 1971. Muscle metabolites during exercise in pubertal boys. *Acta Paediatr. Scand. (Suppl.)* 217: 154-57.

Eriksson, B.O., and B. Saltin. 1974. Muscle metabolism during exercise in boys aged 11 to 16 years compared to adults. *Acta Paediatr. Belg. (Suppl.)* 28: 257-65.

Evans, J.A., and H.A. Quinney. 1981. Determination of resistance settings for anaerobic power testing. *Can. J. Appl. Sport Sci.* 6: 53-56.

Falk, B., and O. Bar-Or. 1993. Longitudinal changes in peak aerobic and anaerobic mechanical power of circumpubertal boys. *Pediatr. Exerc. Sci.* 5: 318-31.

Fournier, M., J. Ricci, A.W. Taylor, R. Ferguson, R. Montpetit, and B. Chaitman. 1982. Skeletal muscle adaptation in adolescent boys: Sprint and endurance training and detraining. *Med. Sci. Sports Exerc.* 14: 453-56.

Fox, E., R. Bartels, J. Klinzing, and K. Ragg. 1977. Metabolic responses to interval training programs of high and low power output. *Med. Sci. Sports Exerc.* 9: 191-96.

Fox, E., and D. Mathews. 1974. *Interval training.* Philadelphia: Saunders.

Geron, E., and O. Inbar. 1980. Motivation and anaerobic performance. In *Art and science of coaching*, ed. U. Simri, 107-17. Netanya, Israel: Wingate Institute.

Grodjinovsky, A., and O. Bar-Or. 1984. Influence of added physical education hours upon anaerobic capacity, adiposity, and grip strength in 12 to 13-year old children enrolled in a sports class. In *Children and sport*, ed. J. Ilmarinen and I. Välimäki, 162-69. New York: Springer Verlag.

Grodjinovsky, A., O. Inbar, R. Dotan, and O. Bar-Or. 1980. Training effect in children on performance as measured by the Wingate Anaerobic Test. In *Children and exercise IX*, ed. K. Berg and B.O. Eriksson, 139-45. Baltimore: University Park Press.

Hamel, P., J.A. Simoneau, G. Lortie, M. Boulay, and C. Bouchard. 1986. Heredity and muscle adaptation to endurance training. *Med. Sci. Sports Exerc.* 18: 690-96.

Hanning, R.M., C.J.R. Blimkie, O. Bar-Or, L.C. Lands, L.A. Moss, and W.M. Wilson. 1993. Relationships among nutritional status and skeletal and respiratory muscle function in cystic fibrosis: Does early dietary supplementation make a difference? *Am. J. Clin. Nutr.* 57: 580-87.

Hebestreit, H., K. Mimura, and O. Bar-Or. 1993. Recovery of anaerobic muscle power following 30-s supramaximal exercise: Comparing boys and men. *J. Appl. Physiol.* 74: 2875-80.

Heiser, K. 1989. Load optimization for peak and mean power output on the Wingate Anaerobic Test. Unpublished master's thesis. Arizona State University.

Hermansen, L. 1969. Anaerobic energy release. *Med. Sci. Sports Exerc.* 1: 32-38.

Hermansen, L., and J. Medbø. 1984. The relative significance of aerobic and anaerobic processes during maximal exercise of short duration. In *Physiological chemistry of training and retraining*, ed. P. Marconnet and J. Poortmans, 56-57. Basel: Karger.

Hill, A.V., C.N.H. Long, and H. Lupton. 1924. Muscular exercise, lactic acid, and the supply and utilization of oxygen. Part VII-VIII. *Proc. Roy. Soc. London* 97: 155-76.

Horswill, C.A., D.L. Costill, W.J. Fink, M.G. Kirwan, J.B. Mitchell, and J.A. Houmard. 1988. Influence of sodium bicarbonate on sprint performance: Relationship to dosage. *Med. Sci. Sports Exerc.* 20: 566-569.

Houston, M., and J. Thomson. 1977. The response of endurance-adapted adults to intense anaerobic training. *Eur. J. Appl. Physiol.* 36: 206-13.

Inbar, O. 1985. *The Wingate Anaerobic Test: Its performance, characteristics, application, and norms.* Netanya, Israel: Wingate Institute (Hebrew).

Inbar, O., D.X. Alvarez, and M.A. Lyons. 1981. Exercise-induced asthma—a comparison between two modes of exercise stress. *Eur. J. Respir. Dis.* 62: 160-67.

Inbar, O., and O. Bar-Or. 1975. The effects of intermittent warm-up on 7-9 year-old boys. *Eur. J. Appl. Physiol.* 34: 81-89.

———. 1977. Relationships of anaerobic and aerobic arm and leg capacities to swimming performance of 8-12 year old children. In *Frontiers of activity and child health,* ed. R.J. Shephard and H. Lavallée, 283-92, Quebec: Pelican.

———. 1980. Changes in arm and leg anaerobic performance in laboratory and field tests following vigorous physical training. *Proceedings of the international seminar on the art and science of coaching,* 38-48. Netanya, Israel: Wingate Institute.

———. 1986. Anaerobic characteristics in male children and adolescents. *Med. Sci. Sports Exerc.* 18: 264-69.

Inbar, O., R. Dotan, and O. Bar-Or. 1976. Aerobic and anaerobic components of a thirty-second supramaximal cycling task. *Med. Sci. Sports Exerc.* 8: S51.

Inbar, O., R. Dotan, T. Trousil, and Z. Dvir. 1983a. The effect of bicycle crank-length variation upon power performance. *Ergonomics* 26: 1139-46.

Inbar, O., M. Epstein, R. Dlin, Y. Weinstein, and A. Kowalski. 1989. Physiological profiling of elite athletes. In *Proceedings of the first IOC World Congress on Sport Sciences,* 74-76. Colorado Springs, CO: U.S. Olympic Committee.

Inbar, O., P. Kaiser, and P. Tesch. 1981. Relationships between leg muscle fiber type distribution and leg exercise performance. *Int. J. Sports Med.* 2: 154-59.

Inbar, O., A. Rotstein, I. Jacobs, P. Kaiser, R. Dlin, and R. Dotan. 1983b. The effects of alkaline treatments on short-term maximal exercise. *J. Sports Sci.* 1: 95-104.

Jacobs, I. 1979. The effects of thermal dehydration on performance of the Wingate test of anaerobic power. Unpublished master's thesis, 1-76. University of Windsor, Windsor, Ontario.

———. 1980. The effects of thermal dehydration on performance of the Wingate Anaerobic Test. *Int. J. Sports Med.* 1: 21-24.

Jacobs, I., O. Bar-Or, J. Karlsson, R. Dotan, P. Tesch, P. Kaiser, and O. Inbar. 1982. Changes in muscle metabolites in females with 30-second exhaustive exercise. *Med. Sci. Sports Exerc.* 14: 457-60.

Jacobs, I., P.A. Tesch, O. Bar-Or, J. Karlsson, and R. Dotan. 1983. Lactate in human skeletal muscle after 10 and 30 seconds of supra-maximal exercise. *J. Appl. Physiol.* 55: 365-67.

Jones, N.L., and N. McCartney. 1986. Influence of muscle power on aerobic performance and the effect of training. *Acta Med. Scand.* Suppl. 711: 115-22.

Kaczkowski, W., D.L. Montgomery, A.W. Taylor, and V. Klissouras. 1982. The relationship between muscle fiber composition and maximal anaerobic power and capacity. *J. Sports Med. Phys. Fit.* 22: 407-13.

Karlsson, J. 1971. Muscle ATP, CP and lactate in submaximal and maximal exercise. In *Muscle metabolism during exercise*, ed. B. Pernow and B. Saltin, 383-93. New York: Plenum Press.

Karlsson, J., and B. Saltin. 1970. Lactate, ATP, and CP in working muscles during exhaustive exercise in man. *J. Appl. Physiol.* 29: 598-602.

Katch, V. 1973. Kinetics of oxygen uptake and recovery for supramaximal work of short duration. *Int. Z. angew. Physiol.* 31: 197-207.

Katch, V., A. Weltman, R. Martin, and L. Gray. 1977. Optimal test characteristics for maximal anaerobic work on the bicycle ergometer. *Res. Q.* 48: 319-26.

Kavanagh, M.J., I. Jacobs, J. Pope, D. Symons, and A. Hermiston. 1986. The effect of hypoxia on performance of the Wingate anaerobic power test. *Can. J. Appl. Sport Sci.* 11: 22P.

Kemper, H., H. Dekker, M. Ootjers, B. Post, J. Snel, P. Splinter, L. Storm-Van Essen, and R. Verschnuur. 1983. Growth and health of teenagers in the Netherlands: Survey of multi-disciplinary longitudinal studies and comparison to recent results of a Dutch study. *Int. J. Sports Med.* 4: 202-14.

Kindermann, W., G. Huber, and J. Keul. 1975. Anaerobic capacity in children and adolescents in comparison with adults (in German). *Sportarzt Sportmed.* 6: 112-15.

Kohrt, W. 1986. Training specificity and the triathlete. Unpublished doctoral dissertation. Arizona State University.

Komi, P., H. Rusko, J. Vos, and V. Vihko. 1977. Anaerobic performance capacity in athletes. *Acta Physiol. Scand.* 100: 107-14.

Krahenbuhl, G., J. Skinner, and W. Kohrt. 1985. Developmental aspects of maximal aerobic power in children. *Exerc. Sport Sci. Rev.* 13: 503-38.

Krotkiewski, M., J.G. Kral, and K. Karlsson. 1980. Effects of castration and testosterone substitution on body composition and muscle metabolism in rats. *Acta Physiol. Scand.* 109: 233-37.

Kurowsky, T. 1977. Anaerobic power of children from ages 9 through 15 years. Unpublished master's thesis. Florida State University.

Lavoie, N., J. Dallaire, D. Barrett, and S. Brayne. 1984. Anaerobic testing using the Wingate and Evans-Quinney protocols with and without toe stirrups. *Can. J. Appl. Sport Sci.* 9: 1-5.

Lopes, J., D. Russell, J. Whiteell, and K.N. Jeejeebhoy. 1982. Skeletal muscle function in malnutrition. *Am. J. Clin. Nutr.* 36: 602-10.

Mainwood, G., and D. Cechetto. 1980. The effect of bicarbonate concentration on fatigue and recovery in isolated rat diaphragm. *Can. J. Physiol. Pharmacol.* 58: 624-32.

Margaria, R., P. Aghemo, and E. Rovelli. 1966. Measurement of muscular power (anaerobic) in man. *J. Appl. Physiol.* 21: 1662-64.

Margaria, R., H.T. Edwards, and D.B. Dill. 1933. The possible mechanism of contracting and paying the oxygen debt and the role of lactic acid in muscular contraction. *Am. J. Physiol.* 106: 689-715.

Margaria, R., D. Oliva, P.E. Di Prampero, and P. Cerretelli. 1969. Energy utilization in intermittent exercise of supramaximal intensity. *J. Appl. Physiol.* 26: 752-56.

Markides, L., G.J.F. Heigenhauser, N. McCartney, and N.L. Jones. 1985. Maximal short-term exercise capacity in healthy subjects aged 15-70 years. *Clin. Sci.* 69: 197-205.

Matejkova, J., Z. Koprivova, and Z. Placheta. 1980. Changes in acid-base balance after maximal exercise. In *Youth and physical activity*, ed. Z. Placheta, 191-99. Brno: J.E. Purkynje University.

Maud, P.J., and B. Shultz. 1989. Norms for the Wingate Anaerobic Test with comparison to another similar test. *Res. Q. Exerc. Sport* 60: 144-51.

McCartney, N., G.J.F. Heigenhauser, and N.L. Jones. 1983. Power output and fatigue of human muscle in maximal cycling exercise. *J. Appl. Physiol.* 55: 218-24.

Morse, M., F.W. Schultz, and D.E. Cassels. 1949. Relation of age to physiological responses of the older boy (10 to 17 years) to exercise. *J. Appl. Physiol.* 1: 683-709.

Murphy, M.M., J.F. Patton, and F.A. Frederick. 1984. A comparison of anaerobic power capacity in males and females accounting for differences in thigh volume, body weight, and lean body mass (abstract). *Med. Sci. Sports Exerc.* 16: 108.

O'Connor, J. 1987. A cross-sectional and longitudinal investigation of a physical activity classification system. Unpublished doctoral dissertation. Arizona State University.

Parker, D.F., L. Carriere, H. Hebestreit, and O. Bar-Or. 1992. Anaerobic endurance and peak muscle power in children with cerebral palsy. *Am. J. Dis. Child.* 146: 1069-73.

Parkhouse, W.S., and D.C. McKenzie. 1984. Possible contribution of skeletal muscle buffers to enhanced anaerobic performance: A brief review. *Med. Sci. Sports Exerc.* 16: 328-38.

Patton, J., M. Murphy, and F. Frederick. 1985. Maximal power outputs during the Wingate Anaerobic Test. *Int. J. Sports Med.* 6: 82-85.

Perez, H.R., J.W. Wygand, A. Kowalski, T.K. Smith, and R.M. Otto. 1986. A comparison of the Wingate power test to bicycle time trial performance. *Med. Sci. Sports Exerc.* 18: S1.

Pirnay, F., and J. Crielaard. 1979. Mesure de la puissance anaérobic alactique. *Med. du Sport* 5: 13-16.

Reilly, T. 1987. Circadian rhythm and exercise. In *Exercise, benefits, limits and adaptations*, ed. D. MacLeod, R. Maughan, N. Nimmo, T. Reilly, and C. Williams, 346-66. London: E. and F.N. Spon.

Reilly, T., and C. Baxter. 1983. Influence of time of day on reactions to cycling at a fixed high intensity. *Br. J. Sports Med.* 17: 128-30.

Reilly, T., and A. Down. 1986. Time of day and performance of all-out arm ergometry. In *Kinanthropometry III*, ed. T. Reilly, J. Watkins, and J. Borms, 296-300. London: E. and F.N. Spon.

Rhodes, E.C., M.H. Cox, and H.A. Quinney. 1986. Physiological monitoring of national hockey league regulars during the 1985-1986 season. *Can. J. Appl. Sport Sci.* 11: 36P.

Rhodes, E.C., R.E. Mosher, and J.E. Potts. 1985. Anaerobic capacity of elite pre-pubertal ice-hockey players. *Med. Sci. Sports Exerc.* 17: S265.

Roberts, A., R. Billeter, and H. Howald. 1982. Anaerobic muscle enzyme changes after interval training. *Int. J. Sports Med.* 3: 8-21.

Robinson, S. 1938. Experimental studies of physical fitness in relation to age. *Int. Z. angew. Physiol. einschl. Arbeitsphysiol.* 10: 251-323.

Russell, D., P.J. Prendergast, P.L. Darsby, P.E. Garfinkel, J. Whitewell, and K.N. Jeejeebhoy. 1983. A comparison between muscle function and body composition in anorexia nervosa: The effect of refeeding. *Am. J. Clin. Nutr.* 38: 229-37.

Rutenfranz, J. 1986. Longitudinal approach to assessing maximal aerobic power during growth: The European experience. *Med. Sci. Sports Exerc.* 18: 270-75.

Saltin, B., and J. Karlsson. 1971. Muscle glycogen utilization during work of different intensities. In *Muscle metabolism during exercise*, ed. B. Pernow and B. Saltin, 289-99. New York: Plenum Press.

Sargeant, A.J. 1989. Short-term muscle power in children and adolescents. In *Advances in pediatric sports sciences*, ed. O. Bar-Or, 42-65. Champaign, IL: Human Kinetics.

Sargeant, A.J., E. Hoinville, and A. Young. 1981. Maximum leg force and power output during short-term dynamic exercise. *J. Appl. Physiol.* 51: 1175-82.

Sargent, D.A. 1921. The physical test of a man. *Am. Phys. Educ. Rev.* 26: 188-94.

Sharp, R.L., P.D. Stanford, L. Bevan, and W.S. Runyan. 1986. Effects of altered acid-base status on performance of two types of anaerobic exercise tests. *Med. Sci. Sports Exerc.* 18: S2.

Shephard, R., C. Allen, O. Bar-Or, C. Davies, S. Degré, R. Hedman, K. Ishii, M. Kaneko, J. LaCour, P.E. Di Prampero, and V. Seliger. 1969. The working capacity of Toronto school children. *Can. Med. Assoc. J.* 100: 560-66, 705-14.

Simoneau, J.A., G. Lortie, M. Boulay, M. Marcotte, M. Thibault, and C. Bouchard. 1986. Inheritance of human skeletal muscle and anaerobic capacity adaptation to high-intensity intermittent training. *Int. J. Sports Med.* 7: 167-71.

Sjödin, B., A. Thorstensson, K. Frith, and J. Karlsson. 1976. Effect of physical training on LDH activity and LDH isozyme pattern in human skeletal muscle. *Acta Physiol. Scand.* 97: 150-57.

Skinner, J.S., and D. Morgan. 1985. Aspects of anaerobic performance. In *Limits of human performance*, ed. D. Clarke and H. Eckert, 31-44. Champaign, IL: Human Kinetics.

Skinner, J.S., and J. O'Connor. 1987. Wingate test: Cross-sectional and longitudinal analysis. *Med. Sci. Sport Exerc.* 19: S73.

Smith, D.J., H.A. Quinney, and R.D. Steadward. 1982. Physiological profiles of the Canadian Olympic hockey team (1980). *Can. J. Appl. Sport Sci.* 7: 142-46.

Stevens, G.H.J., and B.W. Wilson. 1986. Aerobic contribution to the Wingate test. *Med. Sci. Sports Exerc.* 18: S2.

Szögÿ, A., and G. Cherébetiu. 1974. Minutentest auf dem Fahrradergometer zur Bestimmung der anaeroben Kapazität. *Eur. J. Appl. Physiol.* 33: 171-76.

Tamayo, M., A. Sucec, W. Phillips, M. Buon, L. Laubach, and M. Frey. 1984. The Wingate Anaerobic Test, peak blood lactate, and maximal oxygen debt in elite volleyball players: A validation study. *Med. Sci. Sports Exerc.* 16: S126.

Taunton, J.E., H. Maron, and J.G. Wilkinson. 1981. Anaerobic performance in middle and long-distance runners. *Can. J. Appl. Sport Sci.* 6: 109-13.

Tharp G., G. Johnson, and W. Thorland. 1984. Measurement of anaerobic power and capacity in elite young track athletes using the Wingate test. *J. Sports Med.* 24: 100-106.

Tharp, G.D., R.K. Newhouse, L. Uffelman, W.G. Thorland, and G.O. Johnson. 1985. Comparison of sprint and run times with performance on the Wingate Anaerobic Test. *Res. Q. Exerc. Sport* 56: 73-76.

Thompson, N.N., C. Foster, B. Rogowski, and K. Kaplan. 1986. Serial responses of anaerobic muscular performance in competitive athletes. *Med. Sci. Sports Exerc.* 18: S1.

Thorstensson, A., B. Hultén, W. von Döbeln, and J. Karlsson. 1976. Effect of strength training on enzyme activities and fibre characteristics in human skeletal muscle. *Acta Physiol. Scand.* 96: 392-98.

Thorstensson, A., and J. Karlsson. 1976. Fatiguability and fibre composition of human skeletal muscle. *Acta Physiol. Scand.* 98: 318-22.

Thorstensson, A., B. Sjödin, and J. Karlsson. 1975. Enzyme activities and muscle strength after "sprint training" in man. *Acta Physiol. Scand.* 94: 313-18.

Tirosh, E., P. Rosenbaum, and O. Bar-Or. 1990. New muscle power test in neuromuscular disease. Feasibility and reliability. *Am. J. Dis. Children* 144: 1083-87.

Vandewalle, H., G. Peres, J. Heller, and H. Monod. 1985. All-out anaerobic capacity tests on cycle ergometers. *Eur. J. Appl. Physiol.* 54: 222-29.

Vandewalle, H., G. Peres, and H. Monod. 1987. Standard anaerobic exercise tests. *Sports Med.* 4: 268-89.

Van Mil, E., N. Schoeber, R. Calvert, and O. Bar-Or. 1993. Prediction of optimal braking force for the Wingate test during arm cranking in children and adolescents with neuromuscular disabilities. *Pediatr. Exerc. Sci.* 5: 481.

Van Praagh, E., N. Fellmann, M. Bedu, G. Falgairette, and J. Coudert. 1990. Gender difference in the relationship of anaerobic power to body composition in children. *Pediatr. Exerc. Sci.* 2: 336-48.

Volkov, N., E. Shirkovets, and V. Borilkevich. 1975. Assessment of aerobic and anaerobic capacity of athletes in treadmill running tests. *Eur. J. Appl. Physiol.* 34: 121-30.

Von Ditter, H., P. Nowacki, E. Simai, and U. Winkler. 1977. Das Verhalten des Saure-Basen-Haushalts nach erschopfender Belastung bei untrainierten und trainierten Jungen und Mädchen im Vergleich zu Leistungssportlern. *Sportartzt Sportmed.* 28: 45-48.

Watson, R.D., and T.L.C. Sargeant. 1986. Laboratory and on-ice test comparisons of anaerobic power of ice-hockey players. *Can. J. Appl. Sport Sci.* 11: 218-24.

Weltman, A., R. Moffatt, and B. Stamford. 1978. Supramaximal training in females: Effects on anaerobic capacity and aerobic power. *J. Sports Med. Phys. Fit.* 18: 237-44.

Wirth, A., E. Trager, K. Scheele, D. Mayer, K. Diehm, K. Reisckle, and H. Weicker. 1978. Cardiopulmonary adjustment and metabolic response to maximal and submaximal physical exercise of boys and girls at different stages of maturity. *Eur. J. Appl. Physiol.* 39: 229-40.

Index

About the Authors

Omri Inbar, FACSM, EdD, was research associate in the Department of Research and Sports Medicine at the Wingate Institute from 1973 to 1986 and was head of the Life Sciences Department from 1986 to 1991. In 1984 he founded the Exercise Physiology Department at the Mor Institute for Medical Data in Tel Aviv, serving as its director until 1990. He helped establish the exercise medicine unit at the Lincs Clinic in Edmonton, Canada, in 1991 and served as its director of clinical physiology until 1994.

Inbar returned to the Wingate Institute in 1995 and now is scientific consultant to several major Israeli organizations. He is a member of many physiology associations, including the American Physiological Society, the Israeli Society of Physiology and Pharmacology, the Israeli Sports Medicine Association, and the Canadian Society for Exercise Physiology.

Since earning his doctorate in Applied Physiology from Columbia University, Inbar has had articles published in numerous professional books and journals and has lectured at scientific meetings worldwide.

Oded Bar-Or, MD, was head of the Department of Research and Sports Medicine at the Wingate Institute when the Wingate Anaerobic Test was developed. He is a professor of pediatrics and director of the Children's Exercise and Nutrition Centre of McMaster University and Chedoke-McMaster Hospitals in Hamilton, Ontario, where most of the subsequent development and adaptation of the test has taken place.

A speaker at conferences worldwide, Bar-Or has done extensive research on the responses of children and adolescents to exercise. His research has been published in journals such as the *Journal of Applied Physiology, Medicine and Science in Sports and Exercise*, the *American Journal of Clinical Nutrition*, and *Pediatric Exercise Science*.

Bar-Or received his medical degree at the Hebrew University in Jerusalem, Israel, and was the founder and director of the Department of Research and Sports Medicine at the Wingate Institute until assuming his current position in Canada.

He has been president of the Canadian Association for Sports Sciences (now the Canadian Society for Exercise Physiology), president of the International Council for Physical Fitness Research, vice president of the American College of Sports Medicine, and a board member for the Sports Medicine Council of Canada.

In 1996 **James S. Skinner**, PhD, became professor of kinesiology at Indiana University. Prior to the move, he was a professor in the Department of Exercise Science and Physical Education at Arizona State University, where he directed the Exercise and Sport Research Institute (1983-1995). Skinner, and colleagues from four other universities, earned a five-year, $9.5 million research grant from the National Heart, Lung, and Blood Institute.

A past president of the American College of Sports Medicine, Skinner is also a member of the American Alliance of Health, Physical Education, Recreation and Dance, the American Academy of Kinesiology and Physical Education, and numerous other national and international professional organizations. His work has been published in over 100 professional journals, and he has lectured in 37 countries.

Skinner earned his doctorate in Physical Education from the University of Illinois at Urbana-Champaign.